Halophyte Plant Diversity and Public Health

Munir Ozturk • Volkan Altay
Moona Nazish • Mushtaq Ahmad
Muhammad Zafar

Halophyte Plant Diversity and Public Health

Springer

Munir Ozturk
Department of Botany & Centre for Environmental Studies
Ege University
Izmir, Turkiye

Moona Nazish
Department of Plant Sciences
Quaid-i-Azam University
Islamabad, Pakistan

Muhammad Zafar
Department of Plant Sciences
Quaid-i-Azam University
Islamabad, Pakistan

Volkan Altay
Biology Department
Hatay Mustafa Kemal University
Hatay, Turkiye

Mushtaq Ahmad
Department of Plant Sciences
Quaid-i-Azam University
Islamabad, Pakistan

ISBN 978-3-031-21943-6 ISBN 978-3-031-21944-3 (eBook)
https://doi.org/10.1007/978-3-031-21944-3

This Springer imprint is published by the registered company Springer Nature Switzerland AG
The registered company address is: Gewerbestrasse 11, 6330 Cham, Switzerland

Preface

Our planet is burning, and all living beings are under threat. The thresholds like sea level rise and Arctic sea ice loss have already been crossed. The academicians keep on talking about priority areas for global action, but few successful actions have been taken. Today, we see that climate indicators have worsened, in spite of the pandemics. The world population has reached nearly 7 billion, 1/6 of the population of our Earth lives in the arid or/and semi-arid areas. The global problems are expected to come to the forefront because of the activities of living beings living under these conditions. An immediate action is needed because we see droughts, heatwaves, wildfires, storms, and floods. There is need for mitigation of destructive weather events. We must look for acceptable climate restoration before 2050. If we want to save all living beings, we must learn to use resources efficiently, develop and adopt alternative sources so as to get significantly positive results by 2050. The success depends on the stringent protection and restoration of our ecosystems. The carbon-accumulating systems such as forests, wetlands, and grasslands should be protected from degradation. The success in this connection will depend on addressing simultaneously the climate change and biodiversity loss rapidly. The urgent areas for action include nature, food, population, energy, pollutants, and economy. This will be an opportunity for humanity and the resilience. Food production must be arranged in such a way that we succeed to enhance regional self-reliance.

The food systems running successfully depend much on locally available foods, their varieties, as well as traditional and modern systems. Humanity is facing an unprecedented challenge in the globalization of food production and supply. These are a must in the health and well-being of Asian as well as African populations. The simplification of diets could help much in the transitions in food insecurity and nutrition, as these contribute to the emerging epidemics of non-communicable disease. In order to increase regional productivities, for improving food production we have to promote agroecological practices, intensive fertilizer/pesticide use should be reduced, and depleted soils and degraded lands should be restored. The innovations supporting greater well-being on lighter footprints are to be followed. In order to control agricultural emissions, the best way is to use less fertilizers. The timings of application should be organized and smaller amounts used during the annual growth cycle. For biodiversity safety as well as production, very few ecosystems are sustainably managed. The interdependent ecosystem services like pollination, natural flood control, and water purification have suffered seriously and only few ecosystem goods and services are now available to our "biosystems." The human population outburst has destroyed 83% of wild animals and 50% of biomass. In order to protect our intact ecosystems, biodiversity, structural complexity, soils, mycorrhizal associations, plants, lichens, insects, and others, a nature-based solution like reforestation is needed which can protect humanity. If we cross the global aim of protecting an effective land-water percentage by 2050, much of our biodiversity can be restored safely. The global ecosystems will show a stability. Protection as well as restoration of forests as well as other ecosystems will prove highly beneficial. A large-scale conservation is needed for the recovery of natural habitats in order to get enough resilience for supporting the survival and migration of biodiversity, including humans. There is an urgent need for "reduction-removal-repair (restore)." This has to be applied at all levels from our homes to the global level. The habitat transformation needs to be reduced even stopped because big natural solutions are required for restoration of abandoned farmland soils for food systems.

Following much improvement in the nutrition levels, number of undernourished people globally has started to go up again. The droughts are affective and will affect food production at all levels. Reports mention that food production will increase by 2050. However, the current affects are already beyond the levels complying with planetary health goals. An overuse of lands is degrading our soils, the fertilizers will soon become scarce and costly; if our business continues as usual, food production up to 2050 will be insufficient. Approximately 3 billion people can continue to live sustainably with the current production and consumption and a balanced diet. If we move to plant-based dietary ways, these will provide greatest benefits from underutilized food systems. For food security, mitigation of climate change, and safety of our biodiversity, soil protection practices should be followed and prioritized following regenerative agriculture. In Asia and Africa, low yields suggest potential for productivity gains, but the soil needs replenishment of nutrients removed by the crops grown, big irrigation investments, as well as widespread fertilizer use. Most crucial elements of food system transformation are the shifts in production in the form of high-impact foods, e.g., animal products to low impact foods like underutilized crops may prove helpful to a large extent. An urgent action is required across the entire global food system simultaneously on many fronts for a convergence of various resource limits and environmental crises. We get services of immense economic value from Earth's natural capital as well as the biodiversity it supports. These also provide critical sources of food and medicine. Protection and restoration must be the core strategy of economic development and climate action in order to prevent early collapse of ecosystems and dependent economies. This is needed because of the immense ecological and economic returns of protecting what's left and restoring natural capital and ecosystem services as these are the need of the day. The vast ground saline water or seawater resources can be used for the production of economically important crops from the indigenous plants distributed in coastal and inland salt marshes and dry lands.

"Halophytes" are known as plants naturally equipped with the mechanisms to survive under highly saline, arid conditions and produce high biomass. These include salt-tolerant plants: Hygrohalophytes and Hydrohalophytes include plants in salt marshes, Xerohalophytes comprise salt-desert species, Psammohalophytes cover sandy habitat taxa found on littoral or inland sand dunes, and Xerophytes include desert species suspected as halophytes. These ecological types are based on the habitats where they are distributed. Their habitats too are expected to get severely affected by the foreseen climate trends in the future, potentially more acceleration will be observed and negative changes in the distribution and abundance of some endangered halophytes can be expected.

The halophytes can be evaluated as medicinal plants, in the production of edible and essential oils, fodder, and forage because they hold great potential. There is a need to study to development of coastal salt deserts into man-made ecosystems for agricultural productivity. Several medicinal/aromatic plants can be cultivated easily on slightly saline-alkaline soils using seawater irrigation. Many salt-tolerant plants can be domesticated for better economic returns. Halophytes can counteract adverse environmental affects like climate change, marine discharge waters, ecosystem restoration, and the enhancement of primary productivity. It is now well known that conventional agriculture cannot be a solution for food security. Halophytes are naturally equipped with the mechanisms to survive under highly saline and arid conditions and produce high biomass, which could be used as fodder, forage, medicine, edible oil, and in some cases as food for humans. Halophytes are an intriguing ecological group of plants that could be approached for a better understanding, only in an integrated manner, by interdisciplinary and multidisciplinary studies. Each topic has been examined in a comprehensive way, presenting relevant knowledge about halophytes. These plants can contribute to food security and dietary quality. Today, humans are exposed to new health challenges due to ecosystem degradations, dieting simplifications, and species loss, and the knowledge about them is getting lost too. Several underutilized plants are rich in nutrient quality, including foods which reduce the risk of several diseases. In order to improve their livelihoods and offer nutritional and cultural benefits, small-scale farmers can manage and use wild biodiversity.

The information pooled up in this book will demonstrate the relationship between biodiversity and nutrition, convincing decision-makers about the validity of food-based approaches to health. The data available on the mechanisms of action of halophytes give a perspective on their contribution to the health of individuals from biomedical point of view. However, such data are minimal for many species. Currently, an analysis of composition, digestibility, and safety food function holds high priority. The under-investigated wild foods deserve much attention for basic characterization and for ecological, agronomic, and marketing purposes. In order to support the biodiversity use for health purposes, successful interventions are required. These should be considered at multi-sectoral, multidisciplinary levels and must be problem focused. For resolving health problems related to malnutrition, we must first accept the dietary diversity as a fundamental, cost-effective, and sustainable way. For ensuring environmental sustainability, including a reverse in the loss of environmental resources, biodiversity has an important role in addressing the goal of eradicating extreme poverty and hunger together with others related to health.

In Asia, nearly 260 million indigenous communities are living in 17 countries. The bio-cultural treasure of the continent is maintained by these people. This treasure includes foods and medicines of direct importance to human well-being. In addition to medicinal uses, the large number of species that are consumed as part of traditional diets and their contributions to health stand out. The recently emerged pandemic of "Coronavirus" has led to mass mortalities. There are plant-based biomolecules which can be effective alternative medication for various diseases. Herbal medicines can overcome the drawbacks associated with synthetic drugs, in the management of neurodegenerative disorders.

This book is of great significance in developing awareness about the beneficial utilization of halophytes as a potential solution to the food security of people in the salt-effected areas with a focus on the details related to the themes like; introduction covering information on the saline – alkaline areas, morpho-anatomical characteristics of these plants, their floristic, syntaxonomical/ecological features supported by the panoramic views of saline habitats in Chap. 1. In Chap. 2, details on some representative medicinal halophytes have been dealt with as these play an important role as therapeutics. The phytochemistry and biological activity of representative species have been summarized in Chap. 3, which confirms their nutritional and medicinal importance for being a rich source of vitamins, antioxidants, primary/secondary metabolites, minerals, and fibers. The fourth chapter presents ethnic aspects and their importance in the economy. The last chapter includes concluding remarks and future prospects related to the halophytes.

This group of plants is an underutilized one but has the potential to serve as an alternative food source which can reduce pressure on conventional cropping systems, restore the degraded lands, and serve as a source for economically useful therapeutics.

The global society is at a highly critical juncture, and we have to decide how we can reach the goal. We hope the readers will make use of this book for finding innovative and necessary models to achieve food security in dry lands. It discusses important medicinal aspects related to the halophytes, adds to the understanding of halophytes' significance for food security, as well as the economy of dry regions. We hope if this book is evaluated thoroughly by the saline research institutes, industry, and others for a proper utilization of this plant group, it could be a blessing for dry lands together with health as well as food security. We believe this book will further open new perspectives in the research on halophytes. It is expected to support decision-makers globally at different levels in implementing the urgent food systems by 2050.

Izmir, Turkiye	Munir Ozturk
Hatay, Turkiye	Volkan Altay
Islamabad, Pakistan	Moona Nazish
Islamabad, Pakistan	Mushtaq Ahmad
Islamabad, Pakistan	Muhammad Zafar

Acknowledgment

Authors of this book are highly indebted to several colleagues for sharing the original photographs of some halophytes. These are: Prof. Bilquees Gul from Prof. Ajmal Khan Halophyte Centre (Pakistan), Prof. Kristina N. Toderich (Uzbekistan-Japan), Dr. Hadi Pirasteh-Anosheh (Iran), and Prof. I. Faridah-Hanum (Malaysia). From Turkiye, many young colleagues helped us by providing their original photographs, and they are Mustafa Keskin, Aykut Güvensen, Gül Nihan Tuğ, Ahmet Emre Yaprak, and Gül Ayyıldız. Dr. Khafsa Malik (Pakistan) too has helped us in certain aspects.

The motivation from IPBES mail messages inspired us greatly to work on this book. The editors would like to express their special thanks to them.

The success in the preparation of this volume depended largely on the encouragement from the Springer team, who were patient enough to accept our delay in the completion following pandemic. Our greatest appreciation goes to this team.

Contents

About the Authors

Munir Ozturk (PhD, DSc, FIAS, FFPAS) has served as chairman of the Botany Department and founding director of the Centre for Environmental Studies at Ege University, Izmir, Turkiye, including 50 years of teaching and other services in different positions. He has received the fellowships from Alexander von Humboldt, JSPS-Japan, and National Science Foundation of the USA. He has served as a consultant fellow in the Faculty of Forestry at Universiti Putra Malaysia, Malaysia, and as distinguished visiting scientist at the International Centre for Chemical and Biological Sciences, Karachi University, Pakistan. His fields of scientific interest are biosaline agriculture and crops, plant eco-physiology, medicinal and aromatic plants, conservation of plant diversity, as well as pollution and biomonitoring. Dr. Ozturk has published more than 50 books, 80 book chapters, and more than 200 papers in journals with impact factor. He has also been guest editor of 13 journals.

Volkan Altay (PhD, DSc) is a full professor in the Biology Department, Faculty of Science and Arts, Hatay Mustafa Kemal University, Turkiye. His fields of interest are: plant ecology, vegetation ecology, taxonomy, biodiversity, biomonitoring and pollution, and sustainable use of medicinal plant resources. He holds a proven track record of nearly 100 publications in highly reputed professional journals, and has also published 5 books with Springer, nearly 45 research articles in impact factor journals, more than 40 book chapters and reviews, and more than 70 international and national congress reports.

Moona Nazish (PhD) is working in the field of biodiversity. She provided the first draft information on the systematics as well as anatomy, morphology, palynology, adaptations, and structural strategies of halophytes for this book. Dr. Nazish is a member of many national and international science journals/societies. She has published many research articles in national and international journals, and is serving as the reviewer of many well-reputed journals. Her research interests include halophytes, medicinal plants, plant systematics, biodiversity, and pharmaceutical and nutraceutical studies.

Mushtaq Ahmad (PhD) is teaching in the Plant Sciences Department at Quaid-i-Azam University Islamabad. His specialty is plant systematics and biodiversity, and identifying wild plant resources as alternative sources for food, medicine, and biomass energy. He has published over 600 papers in impact factor journals, 25 books, and 17 book chapters. He has been awarded many international and national awards including top 2% influential scientist of the World (2020) including the Best Research Paper Award (HEC 2011), Best Book Award (HEC 2013), and Gold Medal by Pakistan Academy of Sciences (2011). He is member of many international and national academic bodies.

Muhammad Zafar (PhD) is an assistant professor in the Department of Plant Sciences, Quaid-i-Azam University, Islamabad, Pakistan. He is working in the field of medicinal plants, plant biodiversity, and renewable energy. He is also the founder of first nutraceutical and traditional medicinal plants and biofuel/biomass energy laboratories at Quaid-i-Azam University and has published 23 books, several chapters in the field of medicinal plants/biodiversity and energy technology, and 400 research papers in impact factor journals. The Ministry of Science and Technology and Pakistan has awarded him a gold medal and productive scientist award. He is a member of many societies and associate editor of various journals.

Chapter 1
Introduction

1.1 Introduction

We come across diverse ecosystems spread over varying habitats, at different altitudes, and climatic features in the Asian continent, with reasonably good impact on its biodiversity which is representing reasonably rich plant diversity. There are various combinations of biomes differing in their biodiversity which includes thousands of plant taxa and high endemics (Ozturk et al. 2021a, b, 2022; Malik et al. 2021). There are 10 biodiversity spots in the continent out of global 34 international biodiversity hotspots (Malik et al. 2021). An occurrence of 42 plant families points out that 40% of global plants are distributed here, 1500 genera being endemics (Raghunandan and Basavarajappa 2014; Malik et al. 2021).

The number of biomes too is on the whole very rich here, because its landscape shows similar combinations of climate, vegetation, and animal life (Egamberdieva and Ozturk 2018; Ozturk et al. 2021a, b). The major vegetation types are based on the richness and type of each regional flora, showing parallel variation with changes in precipitation and temperature. We find a full spectrum of vegetation from tundra to tropical rainforest, with a typical latitudinal distribution pattern (Majumder et al. 2013; Malik et al. 2021).

The subpolar climate occurs in the north of Asia, covered by tundra vegetation full of mosses, grasses, and other small plants (Park and Sohn 2010; Malik et al. 2021). The tundra gives way to the taiga when one moves further inlands from the Arctic coast. This is followed by vast forested regions of conifers such as spruce, larch, and fir. The taiga mixes with broadleaved tree forests in the south together with a mixed forest of broadleaved as well as needle-leaved trees. In the interior parts of northern areas, the forests merge with vast grasslands, dominated by short, steppe grasses. The semiarid or desert vegetation is mostly covering the larger area

M. Ozturk et al., *Halophyte Plant Diversity and Public Health*,
https://doi.org/10.1007/978-3-031-21944-3_1

of southwest Asia and interior parts. The other vegetation types and short grasses require minimum precipitation, which are surrounded by many barren deserts (Harrison et al. 2001; Malik et al. 2021). The semiarid tropical vegetation is distributed mainly in the eastern parts of South Asia and tropical rain forests dominating the south coastal strips of Sri Lanka (Shuyu and Zhe 2004; Malik et al. 2021). In the coastal strip of the mainland of Southeastern Asia, we come across tropical forests, extending up to south China. The temperate forests are distributed on the north side with chaparral vegetation covering the Bo Hai Gulfs, which is full of up to 5 m tall woody shrubs. Palestine, Israel, Jordan, Lebanon, and Turkiye form the west Asian side which is rich in maquis and phrygana resembling chaparrals (Malik et al. 2021).

The flora generally consists of: trees and shrubs (Altay et al. 2012b; Karahan et al. 2015; Ozyıgıt et al. 2015; Sezer et al. 2015; Altay 2019; Rajpar et al. 2020; Ozturk et al. 2017a, 2021a, b, 2022), orchids (Ozturk et al. 2021a), crop species (Hameed et al. 2017; Iqbal et al. 2018; Kafi et al. 2018; Sardar et al. 2018; Mercimek et al. 2019; Raza et al. 2019; Taşpınar et al. 2019; Mushtaq et al. 2020; Ayub et al. 2021; Haq et al. 2021; Ozturk et al. 2021c; Shahrasbi et al. 2021; Hocaoglu-Ozyigit et al. 2022), garden plants (Osma et al. 2010; Altay 2012; Eskin et al. 2012; Altay et al. 2015a), ferns, bryophytes, and lichens (Ozturk et al., 2018a, b, 2021a, b), grasses, weeds and ruderals (Yarcı et al. 2007; Karahan et al. 2012a, b; Yarcı and Altay 2016; Altay and Karahan 2017b; ur Rehman et al. 2020; Ozturk et al. 2012a, 2021a, 2022; Altay et al. 2010a, b, 2011, 2012a, 2015b, 2020), native taxa (Severoğlu et al. 2011; Tarakçı et al. 2012; Imanberdieva et al. 2018a, b; Ozturk et al. 2021a, 2022), endemics and relicts, some endangered and/or extinct species (Eskin et al. 2013; Eroglu et al. 2014; Altay et al. 2013, 2016a, b, 2017; Ozturk et al. 2020, 2021a, 2022), and medicinal-flowering plants (Altay and Çelik 2011; Altay and Karahan 2012, 2017a; Altay et al. 2015c; Letchamo et al. 2018; Pleskanovskaya et al. 2019; Masoodi et al. 2020; Younessi-Hamzekhanlu et al. 2020; Malik et al. 2021; Ozturk et al. 2012b, c, 2017b, c, d, 2018c, d, e, f, 2021a, 2022).

Saline regions are among the great contributors to biodiversity worldwide. The varied gradients in saline areas require specialized plant species such as halophytes, which represent great diversity of the habitats where they grow and are salt tolerant. Later represent the characteristic vegetation of the "**halobiome,**" an ecologically special situation. The continent under question is a global plant diversity paradise (Malik et al. 2021; Ozturk et al. 2021a, 2022). The halobiome distributed here deserves high priority in conservation practices (Ozturk et al. 2014a, 2016, 2018g, 2019a, b, 2021a; Altay and Ozturk 2012, 2021; Caparrós et al. 2022).

1.2 Saline: Alkaline Areas of Asia and Halophytes

Salinity is the central character of semi-arid and arid environments in the world. It is the major cause of land degradation resulting in an economic loss in such regions (Wang et al. 2018). Sandu (1984) has given the definition of saline soil as "the soils affected by the minerals present in the upper layers of the water table or by the high contents of soluble salts in profile or by the sodium amount in colloidal complex." The salinity of soils is closely related to their aridity, and such regions are increasing day by day due to the weathering of rocks, upper surface evaporation, low precipitation and irrigation with saline waters. Nearly 50% of the globally occurring arid regions are expected to become more saline by 2050 (Shrivastava and Kumar 2015). Globally, many chemical and biological processes and land degradation have resulted in the harmful effects on natural resources during the last century. Salinity is the main cause of threats to natural resources and human health in many countries (Patel et al. 2011).

The area of saline lands in the world is around 6.5 million ha, while 2 million ha are moderately saline. These lands face an important constraint in irrigated agriculture affecting almost 7% of the total land area or more. Currently, more food is required for the our growing population and reclamation of saline lands by making these fertile is not the potential but a need. We have to utilize these resources for the cultivation of cash crop salt-tolerant plants (Flowers and Colmer 2008). We come across such lands in several continents including Asia and large regions in many Asian countries are facing severe salinization such as Pakistan and India (Khan and Qaiser 2006).

Pakistan is number eight in terms of saline lands (Corbishley and Pearce 2007). The soil in the deserts and salt ranges of Punjab are salt-affected due to the presence of sodium and potassium chloride, calcium sulfate and sodium bicarbonate. The species such as *Chenopodium album, Suaeda fruticosa, Salsola kali*, and *Tamarix aphylla* are the most common halophyte taxa distributed on these saline areas, resulting in micro-edaphic arrangements. The coastal areas in the plains of Sindh and Balochistan show a rich diversity of halophytes. The deserts in Punjab and Potwar plateau are also the centers of halophytic plant taxa. The Punjab Salt Range is spread over the Mianwali, Jhelum, Chakwal, Khushab, and Attock districts. The two hill tracts in the salt range run parallel from east to the southwest. We find several salt and brackish water lakes in the Range of Punjab and Sindh such as Khibbiki Lake near Soon valley, Uchalli Lake near Sakesar mountain and Hadero Lake, all showing a variety of halophytes.

In India, out of 9.38 million ha of salt-affected soil 5.5 million ha are saline. Salinity-covered areas in the country are Nicobar and Andaman Islands in the east, Gujarat in the west, Kanyakumari in the south, and Ladakh in the north. The semi-arid and arid regions in Uttar Pradesh, Haryana, Gujarat, and Rajasthan are poor in fresh waters. The extremely salted groundwater is the only water source in these regions.

In Turkiye, the saline areas are represented by the inland and coastal halophytic areas. The coastal saline habitats are located mainly alongside the Aegean, Marmara and Mediterranean Sea coasts. Central Anatolia is the center of inland saline habitats, which shows great halophytic diversity due to the several salt lakes such as Seyfe, Tuz, and Kayseri Lakes.

In the Arabian Peninsula, Gulf has the significant submerged coastal halophytic vegetation. The Aralo-Caspian, Central Iranian, Armenian and Azerbaijani areas linked to Iran, and the Persian Gulf are the four significant distribution habitats for halophytic species in Irano-Turanian regions of Iran. The vast desert regions in the southwest are the center of halophytic flora in Iraq, while in Israel, the saline regions are located in Sede Boqer in the south, Jordan valley in the east, and Negev in the west.

The Huang Ho valleys and Huai and Hai rivers in north China are salt-affected. The saline and arid areas of Afghanistan lie in the southeast and southwest such as Registan, Dashte Margo and Seistan.

Halophytes in the Asian constitute comprise nearly half of the halophytic families and taxa recorded globally. Approximately 1 billion ha of the land in the world is salt-affected but the deserts in Asia constitute a major portion of this area, as it includes different countries, namely, Pakistan, Oman, Iran, Iraq, Israel, Afghanistan, China, India, Jordan, Turkiye, and Yemen. The continent on the whole is mostly semi-arid and arid having sandy and gravel deserts. In Iran, the central plateau is mostly arid. The majority of the deserts are saline with salt flats. Most of the regions of Asia are arid instead of having great resources of water (Ghazanfar et al. 2014). The Euphrates and Tigris rivers flowing through Iraq and Turkiye are important for agriculture in these countries.

Scientists have been studying at length during the last 15–25 years, the distribution of halophytes and their evaluation for food production in semi-arid and arid parts of the world, using salt water for irrigation. The availability of resources of fresh water is scarce in these countries and reserves of salt water are huge (97.47%) (L'vovich 1974). In the 1960s, interesting experiments have been carried out in Israel on the irrigation of plants using seawater. These studies have attracted a great attention of scientists towards the halophytes in saline soil of the semi-arid and arid zones. In view of the global significance of halophytes which has come to the forefront lately, the IAEA has started the research project entitled "Sustainable Use of Saline Groundwater and Wastelands for Plant Growing" in Pakistan, Iraq, Egypt, Germany, Italy, Algeria, Morocco, Jordan, and Syria. These countries are sharing their collected data from the national programs on halophytes. The member countries of this project are utilizing the halophytic sources for the betterment of their economy and to convert the saline wastelands into halophyte cash crop lands utilizing saline waters. Israel is among the largest exporters of ornamental salt-tolerant taxa with an annual profit of $90 million (Shamsutdinov et al. 2017). The faming of halophytes in Asia can help in solving the multidimensional issues such as environmental reclamation, soil fertility, agricultural production, and economy in underdeveloped arid regions (O'Leary 1985). Landscape implimentations using salt tolerant species results in controlled climatic conditions, reduced negative effects on environment and more economic benefits. Arid and hot areas in different countries of Asia with limited potable water resources can earn profit by promoting a national program on the production of ornamental and halophytic crops (Shamsutdinov and Shamsutdinov 2002).

1.3 Morpho-Anatomical Characteristics

Regulation of the tissue-salt concentration relationship in halophytes may cause specific changes at cell, tissue and organ levels through different mechanisms and species-specific morpho-structural and eco-physiological changes (Koyro et al. 2008; Rancic et al. 2019). Adaptation features of the salt-tolerant plants against the current high salinity conditions of the environment are considered as "xeromorphic adaptations" (Rancic et al. 2019). Among these features, the presence of waxy epidermal cuticle and sclerenchyma are accepted as to minimize water loss from plant tissues, the presence of well-developed bulliform cells in leaves, and also the density of trichomes in leaves are some prominent morpho-anatomical features in this connection (Koyro et al. 2008; Flowers et al. 2010; Rancic et al. 2019).

The studies on the halophytes at morphological level reveal that the leaf organ can negatively affect leaf size in proportionality to salinity by providing a water-saving advantage. It has been reported that mangroves develop smaller leaves as salinity increases and stomata density and size, which are critical in controlling water loss in leaves under conditions of salt stress, generally tend to decrease (Omamt et al. 2006; Maricle et al. 2009; Hameed et al. 2010a, b; Reef and Lovelock 2015; Akcin et al. 2017; Rancic et al. 2019). Conversely, where stomatal density is high, the stomata are smaller and embedded in small-sized leaves, if the stomata are not embedded, guard hairs usually predominate on the leaves (Reef and Lovelock 2015; Rancic et al. 2019; Grigore et al. 2014; Grigore 2021).

Existing trichomes, especially on the leaf surfaces of some halophytes, distributed in coastal habitats, have also been observed to be of functional importance such as reducing transpiration as well as providing a protection from sea water droplets (Grigore et al. 2014; Grigore 2021).

The bulliform cells present in halophyte taxa belonging to some families such as Cyperaceae, Juncaceae and Poaceae play a great role in the rolling of leaves and prevent excessive sweating. From an ecophysiological point of view, this provides an advantage of being able to cope with the physiological drought caused by the salinity in the xeromorphic leaves (Stevanovic et al. 2010; Grigore et al. 2014; Grigore 2021).

In the investigations carried out by different researchers, high salinity in terms of morpho-anatomical features often causes significant changes in the structure of a leaf such as organization and distance between vascular bundles, diameter, and number of xylem veins (Reef and Lovelock 2015; Al-Taisan and Gabr 2017; Akcin et al. 2017). It may also cause a decrease in xylem vessel diameter, especially with increasing salinity levels (Akcin et al. 2017).

Another morpho-anatomical feature seen in halophytes is the presence of mucilage, which represents a special case conferring the plant's an ability to overcome physiological drought in existing environments, particularly on high saline habitats. This feature increases the tissue capacity of halophytes due to their hydrophilic properties, allowing more water to be stored in their body (Clarke et al. 1979; Ghanem et al. 2010). Many studies by different researchers have reported that the mucilage content increases due to salt stress in the vegetative parts of halophytes such as *Kosteletzkya virginica*, *Limonium* spp., *Salicornia brachiata* and *Suaeda maritima* (Batanouny et al. 1992; Ghanem et al. 2010; Jaiswar and Kazi 2016).

Although it generally varies according to the ecological conditions of the environment and the plant species, the salt content in the vegetative organs of halophytes increases with aging and can reach toxic levels. Therefore, the salt-tolerant species tend to increase the water content in their tissues and/or decrease the salt concentration (Dajic 2006). Considering these features, one of the adaptive mechanisms frequently seen in halophytes is leaf and stem succulence. This represents the dilution mechanism in halophytes and provides an advantage in reducing the negative osmotic and toxic effects of salt ions (Waisel 1972; Flowers and Colmer 2008; Hameed et al. 2010a, b; Rancic et al. 2019). The succulence present in some characteristic succulent halophytes thriving in the arid environments is induced by the deposition of salts in the cytoplasm, leading to an increase in the osmotic concentration of the cell sap and allowing these plants to easily absorb the saline water. The concentration of salt ions in the cell sap can be much higher than in a soil solution. In addition, water storage allows for greater water uptake and dilution of salts accumulated in the vacuole, which covers more than 70% of the parenchymal cell (Hajibagheri et al. 1984; Youssef 2009; Hameed et al. 2010a, b; Rancic et al. 2019).

Apart from the strategy of succulent halophytes to dilute excess salts in photosynthetic tissue, another group of salt-tolerant species has evolved the strategy of salt excretion by specialized structures of the aerial organs, and this plays an important role in removing salts from buds and young stems (Rancic et al. 2019). Such an excretion is the mechanism typical for plant taxa which develope special features, mostly localized at the leaf and epidermis-the salt glands, salt hairs or salt bladders (Fahn 1988; Rancic et al. 2019). These structures are a key trait of salt-excreting plants, the so-called crynohalophytes. Such halophytes are represented globally by nearly 370 species (Yuan et al. 2016). Excess salt is released through the leaf cuticle, by dissolving in the guttation fluid or by exclusion via special secretory structures (Stenlid 1956; Rancic et al. 2019). About half of all halophytes possess salt bladders or salt glands (Flowers et al. 2010; Rancic et al. 2019; Grigore 2021), which shift ions from leaf tissues to leaf surfaces, where saline solutions crystallize and can be blown away by wind or washed away by rain (Rancic et al. 2019). In some halophyte taxa, the crystallized salts remaining on the leaf surface provide a reflective coating protecting the leaves from strong sunlight; sharp salt crystals also provide an important advantage in terms of serving as a mechanical deterrent for herbivores (Dickison 2000; Rancic et al. 2019).

Salinity also induces changes in root morphology and anatomy (Hameed et al. 2010a, b). Halophytes usually develop many shallow roots. If growing in muddy or loose sandy soil, like some mangrove species, they can develop many stilt or prop roots from the aerial branches of the stem, or a large number of adventitious root buttresses from the basal parts of tree trunks, which provide sufficient support to the plants. Mangrove halophytes have stilt roots developing from stem branches or underground roots, necessary for efficient anchorage in muddy or loose sandy soil (Hameed et al. 2010a, b; Rancic et al. 2019; Grigore 2021).

The root of a typical halophyte has a well-developed cork and a lacunar primary cortex with sclereids and secretory cells and well-developed mechanical and vascular tissue. The anatomical characteristics of the roots of saline desert plants is a reduced cortex, which shortens the distance between the epidermis and vascular bundle in the central part of the root (Dajic 2006). In the roots of halophytes, the endodermis represents a significant barrier controlling salt penetration into the plant body. It plays a crucial role in salt exclusion from the inner tissues, suberization of the exodermis and endodermis is also evident (Dajic 2006; Ding et al. 2010; Krishnamurthy et al. 2014; Reef and Lovelock 2015; Rancic et al. 2019). It has been shown that the Casparian strip of the root endodermis of halophytes is usually much wider (Schreiber et al. 1999; Grigore and Toma 2010; Shabala and Mackay 2011; Rancic et al. 2019; Grigore 2021).

Aerenchyma is well expressed, especially in the underground vegetative organs such as rhizomes and roots. The aerenchyma tissue is found in halophytes in relation to water logging or flooded saline habitats (Flowers and Colmer 2008; Hameed et al. 2010a, b; Rancic et al. 2019), where the plants are subjected to hypoxia or even anoxia and have to store oxygen in the root aerenchyma (Schulze et al. 2005; Rancic et al. 2019). The soil in coastal regions is low in oxygen and to compensate for the lack of soil aeration, plants develop negatively geotropic breathing roots called pneumatophores. The cortex of pneumatophores is made of aerenchyma with large air chambers which are in contact with numerous lenticels on the root surface. These "breathing pores" enable gaseous exchange in poorly aerated soil, and prominent aerenchyma enclosing large air cavities inside the roots help in the conduction of air down to the subterranean or submerged roots (Reef and Lovelock 2015; Rancic et al. 2019).

Palynological studies too are important as these reveal the taxonomic status of flowering plants (Ozturk et al. 2013, 2014b; Altay et al. 2018; Bahadur et al. 2018; Khan et al. 2018). Therefore, such studies are needed to detect and distinguish the taxonomic positions of halophytes at species and genus level. The palynological evidence plays an important role in distinguishing and identifying closely related taxa by providing valuable information about them (Quamar et al. 2018).

1.4 Floristic, Syntaxonomical, and Ecological Features

1.4.1 Floristic Features

Saline soils play an important role as significant ecological entities in many semi-arid and arid landscapes. Halophytes have been able to perform function medicinally as well as a great feed due to their morphological, physiological and ecological characteristics. Such plants are specifically adapted to the environmental limiting factors that determine their geographical distribution in arid regions (Walter 1974).

Plants growing naturally in arid and semi-arid areas are affected by salt and are therefore generally grouped by ecologists as halophytes. These plants are found in arid and semi-arid areas, although common in coastal habitats, tidal lands, salt marshes, inland salt lakes and salt deserts, varying according to the ecological characteristics of environment (Flowers and Colmer 2015; Stevanovic et al. 2019). In addition, the saline habitat plant diversity exhibits different ecological abilities at the molecular, cellular and tissue levels according to their eco-physiological characteristics. This situation has caused each plant to develop its own salt tolerance mechanism (Stevanovic et al. 2019). The special salt-tolerant plants, which represent about 2% of the terrestrial plant species, form an important halophytic flora and vegetation structure of saline habitats (Stevanovic et al. 2019).

In some characteristic arid and semi-arid regions where terrestrial climatic conditions prevail, different salinity types emerge, which vary depending on the geological structure and soil properties of the existing area, depending on the ecosystem formation processes. The best example of this situation is Central Asia, which is an important biodiversity center for halophytes (Wucherer et al. 2001; Breckle et al. 1998, 2001; Breckle 2021). Breckle (2021) has reported that the fourth largest lake on Earth started to shrink since 1960. The same author emphasized that today only small remnants of water bodies are left. The desiccated seafloor is a huge new desert: the Aralkum. Most parts of the Aralkum are saline; thus the region is rich in halophytes which have migrated from the adjacent areas. The new flora of the Aralkum is now about 421 species. About 201 species of these taxa show halophytic features (Breckle 2021).

The majority of the halophytic plants present in Asia belong to the families Asteraceae, Amaranthaceae, Cyperaceae, and Poaceae. According to Ghazanfar et al. (2014), Turkiye has the largest number of halophytic taxa followed by Pakistan, Afghanistan, Yemen, Saudi Arabia, and Oman. The regions of Irano-Turanian including the western Pakistan and Persian Gulf areas are thought to be the diversity center of several halophytic Amaranthaceae genera, namely, *Caroxylon, Salsola, Suaeda, Halanthium, Gamanthus*, and *Kalidium* (Akhani 2004).

In the Arabian Peninsula nearly 120 taxa are reported as halophytes, constituting nearly 4% of the total flora of the Arabian Peninsula (Ghazanfar et al. 2014). In the southwest Asia, the halophytes represent almost half the number of all taxa (and families) recorded for the world by Aronson (1989). In the Arabian Peninsula, the majority of halophytes belong to the families Amaranthaceae, Poaceae,

Zygophyllaceae, Fabaceae, and Plumbaginaceae (Ghazanfar et al. 2019; Ozturk et al. 2019a).

High salt-tolerant plants are among the possible options to use salt-containing soil economically. National development programs with institutional projects need to be developed in rural areas of Asian countries for the economic use of saline lands. In Pakistan from 2001 to 2003 WAPDA has carried out a survey on saline lands and salinity and reported highly saline (7%), medium saline (4%), and slightly saline (10%) lands out of the total land surveyed. A total number of 410 halophytes have been reported by Khan and Qaiser (2006). Out of which 274 have great potential utilization at an economical level. Kokab and Ahmad (2010) also characterized the ecosystem and potential economical utilization of halophytes in Pakistan. *Prosopis Cineraria, Aeluropus lagopoides, Tamarix alii*, and *Salvadora oleoides* are the commonly found halophytes in the deserts of Pakistan.

Ozturk et al. (2014a) have reported that a total of 304 halophytic taxa from 50 families belonging to 172 genera are distributed in Eastern Mediterranean countries (especially Turkiye, Syria, Lebanon, Palestine, and Jordan), which is a typical transition zone between the Saharo-Arabian desert biome and temperate climates. The families with the highest number of taxa found here are Chenopodiaceae/Amaranthaceae (51 taxa), Poaceae (44 taxa), Asteraceae (23 taxa), Fabaceae (22 taxa), Cyperaceae (17 taxa), and Plumbaginaceae (13 taxa), respectively. The genera with the highest number of species are *Limonium, Juncus, Atriplex, Salsola, Suaeda, Tamarix*, and *Plantago*. Also, out of 304 taxa 9 are endemics, and 24 are rare plants (Ozturk et al. 2014a).

The saline areas in Turkiye mainly include salt marshes, salt steppes, and salt meadows. Based on occurrence, these can be separated into coastal and terrestrial areas (Tug et al. 2019). In total, 420 halophytic taxa have been recorded in Turkiye and out of these 65 are endemics (Yaprak and Tug 2006; Ozturk et al. 2008, 2016). A total of 47 endemic taxa are distributed around Tuz Lake (Tug et al. 2011). *Salicornia frietagii, Salsola stenoptera, Lepidium cartilegineum* subsp. *caespitosum*, and *Petrosimonia nigdeensis* occur in almost all halophytic habitats of Central Anatolia. A high level of endemism is especially seen around the southern part of Tuz Lake, being rich in local endemics as well for example *Kalidium wagenitzii, Hypericum salsugineum, Acantholimon halophila*, and *Verbascum pyroliforme*. A higher endemic ratio is recorded in the terrestrial salt marshes, compared to the coastal ones. This is mainly the result of isolation of these areas after the receding of the Tethys Sea (Tug et al. 2019).

The salt marshes and steppes of Eastern Anatolia in Turkiye have similar species composition to the salt steppes and marshes of North Iranian area. *Atriplex portulacoides* and *Salicornia emerici* are common coastal salt-tolerant plants in Turkiye. *Artemisia santonicum* characterizes the salt marshes of Central Anatolia. In Arabian Peninsula, the majority of salt-tolerant plants belong to the families of Fabaceae, Poaceae, Amaranthaceae, and Zygophyllaceae.

Iran has the great diversity of halophytic flora. Akhani (2006) has reported that nearly 365 species belonging to 151 genera and 44 families are distributed in the saltine habitats in Iran. Out of these Chenopodiaceae/Amaranthaceae are represented by 139 species, Poaceae 35 species, Tamaricaceae 29 species, Asteraceae 23 species, Zygophyllaceae 12 species, and Brassicaceae and Juncaceae, each 10 species (Akhani 2006).

Movsumova and Ibadullayeva (2019) have reported that 287 halophytic species of vascular plants are distributed in the flora of Duzdag area in Nakhchivan Autonomous Republic (Azerbaijan). The families here are dominated by Chenopodiaceae/Amaranthaceae, Asteraceae, Fabaceae, Brassicaceae, Poaceae, Apiaceae, and Lamiaceae (Movsumova and Ibadullayeva 2019).

In Afghanistan, *Salsola rosmarinus* is the commonly found halophyte in saline lands. *Halocnemum strobilaceum* is the dominant salt tolerant plant of Afghanistan, North Iran and Central Asia. *Aeluropus lagopoides, Anabasis, Alhagi, Haloxylon, Limonium,* and *Salsola* are other commonly found species (Breckle 2016).

Most of the halophytes belong to the family Amaranthaceae which are dominant in the halophytic flora of Iraq. *Haloxyletum salicornici* is the widespread halophyte of the deserts of Iraq. *Peganum harmala* and *Tamarix passerinoides* are the commonly halophytes in the saline habitats of Iraq (Ghazanfar 2006a).

According to Saenger et al. (2019), approximately 7% of the global mangrove area lies in the South Asia (also known as the Indian subcontinent), which hosts 42 species belonging to 22 genera; maximum number of species occurring in the mainland India (37 species), with a marked decline in numbers with increase in distance from the continental land mass (e.g., Sri Lanka, the Maldives, and the Chagos Archipelago) (Saenger et al. 2019).

The information on the number of halophytic taxa and areas of salt-affected soils present in some Asian countries are summarized in Table 1.1.

1.4.2 Syntaxonomical Features

The high osmotic pressure, ion toxicity, unfavorable soil structure, suboptimal soil pH and nutrient deficiency are the major factors affecting the halophyte vegetation (Füzy et al. 2010; Stevanovic et al. 2019). The vegetation composition in the saline habitat communities is highly influenced by such factors as soil pH and elevation (Wanner et al. 2014; Stevanovic et al. 2019). The vegetation in such habitats is very diverse, complex, fragile, and changing. The vegetational development here too is affected by several notable important factors such as changes in the intensity of salinity and moisture. There are several controversial concepts in halophytic vegetation classification inspite of many years of research. These depend on the classification criteria, geographical scale, and research "school" (Stevanovic et al. 2019).

According to Vuckovic (1982) the local soil conditions and hydrological characteristics are the factors responsible for the inland halophytic vegetation which represents a relevant type of intra-zonal vegetation. We come across high landscape-scale patchiness and rapid shifts among vegetation types in the inland alkaline and saline vegetation due to uneven salt accumulation in the soil and water supply. As per

Table 1.1 Number of halophytic taxa present and areas of salt-affected soils in some Asian countries

Country	Native vascular plants	Number of halophytic taxa	Salt-affected soil areas (x 1000 ha)
China	31,362	**587**	36,658
Asian Russia	6961	**510**	170,720
Turkiye	11,000	**420**	352
Pakistan	6000	**410**	10,456
Iran	8000	**365**	27,085
Palestine	2076	**300**	28
Jordan	2531	**263**	180
Saudia Arabia	2282	**250**	6002
Iraq	3300	**135**	6726
Yemen	2838	**120**	483
Oman	1239	**120**	290
Syria	4200	**110**	532
Bahrain	357	**97**	6
Afghanistan	5000	**96**	3103
Kuwait	407	**80**	209
United Arab Emirates	650	**70**	1089
Qatar	379	**70**	225

Modified from Yensen (2008), Shamsutdinov (2008), Zhao et al. (2011), Nanduri et al. (2019), Ozturk et al. (2019a), Malik et al. (2021)

Kelemen et al. (2013) and Stevanovic et al. (2019), we observe several tolerance responses of plant species to salt stress and different human disturbances.

The coastal vegetation in general comprises of marine algal vegetation in the littoral and sub-littoral zones, while the salt and brackish marshes are covered by the phanerogamic and algal vegetation. According to Stevanovic et al. (2019), the vegetation of sand dunes as well as their "slacks" is a specialized vegetation associated with the drift line; on the shingle beaches the plants occupy coastal cliffs and mangrove vegetation is met with. The regular flooding at high tide is responsible for the saline soils on the seashore where salt concentration decreases proportionally to the distance from the waterline. In accordance with this halophytic species associations differ in degree of salinity tolerance. These are arranged as strips running parallel to the seashore; exceptions being the concentric vegetation belts at the shores of salt- or brackish-water lagoons cut off from the open sea by maritime dunes, which are relatively similar to the vegetation zones around ordinary lakes (Borhidi et al. 2012; Stevanovic et al. 2019).

The prominent halophytic communities and main vegetation types in some Asian countries are presented in Table 1.2.

Table 1.2 The halophytic communities and main vegetation types in some Asian countries

Countries – vegetation types/plant communities – references
Country: CHINA (Chia-Chi and Tsen-Li 1993)
Mangrove communities
Bruguiera formation
Rhizophora formation
Kandelia formation
Aegiceras formation
Avicennia formation
Sonneratia formation
Nypa formation
Salt bush communities
Tamarix chinensis formation
Tamarix ramosissima formation
Tamarix hispida formation
Halimodendron halodendron formation
Nitraria tangutorum formation
Nitraria sibirica formation
Salt desert communities
Reaumuria soongorica community
Ceratoides latens community
Salsola passerina community
Salsola abrotanoides community
Sympegma regelii community
Ilijinia regelii community
Nanophyton erinaceum community
Anabasis aphylla community
Anabasis salsa community
Anabasis brevifolia community
Saline steppe communities
Aneurolepidium chinese community
Saline meadow communities
Saline meadow of rosette grass
Achnantherum splendens formation
Puccinellia tenuiflora and *P. distans* formation
Hordeum brevisubulatum formation
Saline meadow of rhizome grass
Aneurolepidium dasystachys formation
Achnantherum hookeri formation
Aeluropus littoralis var. *sinensis* and *A. littoralis* formation
Phragmites communis formation
Saline meadow of Cypereous grass
Fimbristylis sericea, and *Calystegia soldanella* formation
Saline meadow of mixed grasses

(continued)

Table 1.2 (continued)

Iris lacta var. *chinensis* formation
Sophora alopecuroides formation
Trachomitum lancifolium and *Poacynum hendersonii* formation
Glycyrrhiza injlata formation
Alhagi sparsifolia formation
Karelinia caspica formation
Potentilla anserin formation
Polygonum sibiricum var. *thomsonii* formation
Saline meadow with annual plant communities
Crypsis aculeata community
Suaeda heteroptera community
S. glauca community
S. corniculata community
S. linifolia community
Salsola ruthenica community
Salicornia herbacea community
Salsala ruthenica community
Scorzonera mongolica var. *putnatae* community
Limonium aureum community
Country: TURKIYE (Tug et al. 2019)
Coastal salt marshes associations
Aeluropetum littoralis
Arthrocnemo glauci – Halocnemetum strobilacei
Arthrocnemo macrostachyi–Halocnemetum cruciati
Bolboschoenus maritimus community
Bupleuro tenuissimi-Inuletum viscosae
Carici extensae-Halocnemetum strobilacei
Cressetum cretica
Cresso creticae-Hordeetum marinae
Elytrigio junceae-Zygophylletum albi
Halimiono-Artemisietum santonici
Halimiono portulacoidis-Junceteum littoralis
Halimione portulacoides community
Juncetum maritimum
Juncetum subulati
Junco littoralis-Tamaricetum parviflorae
Limonio gmelinii-Aeluropetum littoralis
Limonio gmelinii-Juncetum gerardii
Limonio gmelinii-Junceteum littoralis
Plantagini weldenii-Parapholidetum incurvae
Phragmito australis-Juncetum maritimii
Phragmites australis community
Puccinellio festuciformis-Arthrocnemetum perennis

(continued)

Table 1.2 (continued)

Puccinellio festuciformis-Caricetum extensae
Puccinellio festuformis-Hordeetum marini
Puccinellio festuciformis-Sarcocornietum fruticosae
Ranunculo marginati-Caricetum divisae
Salicornietum ramosissimae
Sarcocornion fruticosae
Spergulario marinae-Halimionetum portulacoidis
Suaedo-Kochietum hirsutae
Suaedo-Salsoletum sodae
Suaedo-Salicornietum patulae
Tamaricetum smyrnensis
Zygophyllo albi–Halocnemetum cruciati
Terrestrial salt marshes associations
Alopecuretum arundinacea
Bupleuro gracili-Limonietum gmelini
Cardopatio-Juncetum heldreichianii
Cresso creticae-Halocnemetum strobilacei
Eragrostio collinae-Puccinellietum anatolicae
Halocnemetum strobilacei
Inulo aucheranae-Elymetum salsi
Juncetum maritimimum
Lepidio casepitosi-Halimionetum portulacoides
Lepidio caespitosi-Limonietum iconici
Limonio tamaricoidis-Puccinellietum convolutae
Petrosimonietum nigdeenso-brachiatae
Pucciniellio distantis-Juncetum maritimi
Puccinellietum giganteae
Scorzoneretum parviflorae
Sphenopodo divaricati-Halaocnemetum strobilacei
Suaedo anatolicae-Salsoletum nitrariae
Tamaricetum parviflorae-tetrandrae
Salt steppes associations
Achilleo wilhelmsii-Artemisietum santonici
Artemisio scopariae-Peganetum harmalae
Halothamno hierochunticae- Salsoletum incanescentis
Hymenolobo procumbentis-Aeluropetum lagopoidis
Country: KAZAKHSTAN / UZBEKISTAN (Desert Aralkum) (Breckle 2021)
Therophytic plant communities
Salicornia europaea community
Suaeda species (S. acuminata and S. crassifolia) communities
Tripolium vulgare community
Perennial plant communities
Haloxylon aphyllum community

(continued)

Table 1.2 (continued)

Halocnemum strobilaceum community
Kalidium caspicum community
Limonium suffruticosum community
Open salt deserts communities
Suaeda acuminata community
Petrosimonia triandra community
Climacoptera aralensis community
Salt meadow communities
Phragmites australis-Puccinellia dolicholepis community
Limonium otolepsis-Aeluropus littoralis community
Country: IRAN (Akhani 2006)
Halophytic communities
Juncus maritimus-Aeluropus littoralis community
Halanthium rarifolium community
Suaede arcuata community
Anabasis setifera community
Gamanthus pilosus community
Aeluropus littoralis community
Tamarix ramosissima-Aeluropus littoralis community
Salicornia europaea-Suaeda salsa community
Seidlitzia rosmarinus community
Halocnemum strobilaceum-Halostachys belangeriana community
Salicornia persica-Salicornia europaea community
Seidlitzia rosmarinus and Anabasis setifera community
Suaeda cochlearifolia community
Anabasis haussknechtii, Artemisia sieberi, Launaea acanthodes community
Salsola dendroides, Acroptilon repens, Artemisia fragrans community
Halocnemum strobilaceum, Tamarix aucheriana community
Haloxylon ammodendron, Artemisia sieberi community
Halopeplis pygmaea, Cressa cretica community
Halanthium rarifolium, Halothamnus auriculus community
Anabasis haussknechtii, Salsola tomentosa community
Tamarix, Suaede linifolia, Atriplex verrucifera community
Salsola drummondii, Halocharis sulphurea, Halocnemum strobilaceum community
Tamarix, Climacoptera turcomanica community
Tamarix mascatensis, T. kermanensis, Bienertia sinuspersici community
Country: INDIA (Dagar 1995)
Mangrove vegetation (mangal formation)
Excoecaria agallocha community
Avicennia spp. community
Sonneratia apetala community
Ceriops decandra community
Littoral woodlands

(continued)

Table 1.2 (continued)

Casuarina equisetifolia community
Coastal salt marshes
Suaeda fruticosa community
Atriplex stocksii community
Salicornia brachiata community
Urochondra setulosa community
Saline sand formations along beaches
Ipomoea pes-caprae community
Inland salt marshes
Suaeda fruticosa-Aeluropus lagopoides community
Haloxylon salicornicum-Salsola baryosma-Sporobolus marginatus community
Sporobolus halvolus-Cyperus rotundus community
Peganum harmala-Eleusine compressa-Sporobolus marginatus community
Country: IRAQ (Ghazanfar 2006a)
Desert and sabkha vegetation
Haloxyletum ammodendri community
Haloxyletum salicornici community
Zygophylletum coccinei community
Seidlitzietum rosmarini community
Halocnemum strobilacei community
Bienertietum cyclopterae community
Country: SAUDI ARABIA (El-Sheikh and Youssef 1981; El-Sheikh et al. 1985; El-Shourbagy et al. 1987; Mandaville 1990; Böer and Warnken 1992; Böer 1994, 1996; Shaltout et al. 1997; Chaudhary 1998)
Coastal salt marshes
Suaeda monoica, S. fruticosa, and *S. vermiculata* communities
Freshwater communities
Tamarix nilotica and *Salvadora persica* communities
Littoral salt marsh communities
Aeluropus massauensis community
Avicennia marina community
Halopeplis perfoliata community
Limonium axillare-Suaeda pruinosa community
Rhizophora mucronata community
Tetraena coccineum community
Inland sandy saline plains
Caroxylon spp. and *Hammada salicornica*, associated with *Acacia tortilis*
Sabkhas vegetation
Seidlitzia rosmarinus community
Tetraena decumbens-Caroxylon imbricatum community
Country: YEMEN (El-Demerdash et al. 1995; Kürschner et al. 1998; Al-Gifri and Gabali 2002; Al Khulaidi et al. 2010; Al Khulaidi 2013)

(continued)

Table 1.2 (continued)

Coastal vegetation
Avicennia marina community
Suaeda vermiculata community
Limonium cylindrifolium-Suaeda fruticosa-Limonium axillare community
Dunes and flat sandy areas
Cyperus conglomeratus/C. aucheri associations
Odyssea mucronata-Suaeda monoica community
Odyssea mucronata-Panicum turgidum community
Sandy-salty depression communities
Urochondra setulosa association, with the codominant *Arthrophytum macrostachyum, Limonium cylindrifolium*, and *Crotalaria saltiana*
Clayey-salty, relatively wet areas communities
Arthrophytum macrostachyum community
Sandy coastal plain communities
Anabasis ehrenbergii-Pulicaria hadramautica-Tetraena hamiense association
Karstic limestone plateau communities
Stipagrostis paradise, Commiphora gileadensis, and *Euphorbia rubriseminalis*
Coastal dunes
Cyperus aucheri, Coelachyrum piercei, Halopyrum mucronatum, Odyssea mucronata, and *Panicum turgidum*
Country: OMAN (Kürschner 1986; Heathcote and King 1998; Ghazanfar 1992, 1993, 1995, 1998, 1999, 2002, 2006b; and Ghazanfar and Rappenhöner 1994)
Coastal vegetation
Avicennia marina community
Atriplex-Suaeda community
Limonium axillare-Sporobolus-Urochondra community
Limonium sarcophyllum-Suaeda aegyptiaca community
Limonium stocksii-Tetraena qatarense community
Phragmites australis and *Typha* spp. communities
Low coastal dunes which receive salt spray
Atriplex farinosa and *Suaeda moschata* community
Stabilized dunes
Halopyrum mucronatum-Urochondra setulosa community
Saline, silt plains
Arthrocnemum macrostachyum-Suaeda vermiculata community
Shallow sands
Limonium stocksii-Cyperus auheri-Sphaerocoma aucheri community
Country: UNITED ARAB EMIRATES (Deil and Müller-Hohenstein 1996; Böer 2002; Böer and Saenger 2006; Brown et al. 2008)
Coastal dunes and sabkhas vegetation
Avicennia marina community
Arthrocnemum macrostachyum community
Cornulaca monacantha-Sphaerocoma aucheri community

(continued)

Table 1.2 (continued)

Frankenia pulverulenta-Tetraena simplex community
Halopyrum mucronatum, *Atriplex leucoclada*, and *Suaeda aegyptiaca* community
Halopeplis perfoliata community
Halocnemum strobilaceum community
Juncus rigidus-Aeluropus lagopoides community
Limonium axillare community
Salicornia perennans community
Seidlitzia rosmarinus community
Tetraena qatarense community
Country: BAHRAIN (Abbas and El-Oqlah 1992; Abbas 2002)
Sabkhas and coastal lowland commnities
Aeluropus lagopoides community
Avicennia marina community
Halopeplis perfoliata community
Sporobolus ioclados community
Tetraena qatarense community
Country: QATAR (Batanouny 1981; Batanouny and Turki 1983; Babikir 1984; Abdel-Razik and Ismail 1990; Abdel-Razik 1991; Babikir and Kürschner 1992; Böer and Al Hajiri 2002)
Coastal littoral plain communities
Aeluropus lagopoides-Tamarix passerinoides community
Anabasis setifera community
Arthrocnemum macrostachyum community
Avicennia marina community
Caroxylon cyclophyllum-Panicum turgidum-Anabasis setifera community
Halopyrum mucronatum-Sporobolus consimilis community
Halocnemum strobilaceum community
Halopeplis perfoliata community
Juncus rigidus-Aeluropus lagopoides community
Limonium axillare, Suaeda vermiculata, and *Cistanche tubulosa* community
Panicum turgidum and *Pennisetum divisum* community
Salicornia perennans-Suaeda maritima association
Suaeda vermiculata community
Tetraena qatarense community
Country: KUWAIT (Halwagy and Halwagy 1977; Halwagy et al. 1982; Halwagy 1986; Omar et al. 2002; Omar 2007)
Coastal vegetation
Salicornia perennans community
Halocnemum strobilaceum community
Seidlitzia rosmarinus community
Nitraria retusa community
Tetraena qatarense community

1.4.3 Ecological Features

1.4.3.1 Ecological Groupings of Halophytes

The data published by Liang et al. (2017) and Stevanovic et al. (2019) points out that native flora of saline habitats are the "Halophytes." They grow successfully in conditions of 2–6% salts, withstanding even upto 20% soil salt. According to Hasegawa et al. (2000), Dajic (2006), and Stevanovic et al. (2019) the most typical features of the saline habitats are the salt-tolerant plants represented by several different species occupying these habitats and these have evolved a range of adaptive strategies and mechanisms to survive the different levels of salt stress.

Based on salt tolerance, we divide the plants as "**euhalophytes** (tolerant)" and "**glycohalophytes** (sensitive)," with several transitional types such as salt-excreting plants and salt-avoiding plants, for example as stressed by Voronkova et al. (2008) and Stevanovic et al. (2019). Just within the same habitat we come across several different types of salt-tolerant plants, distributed spatially and temporally on local and specific environmental conditions. The conditions can vary on a very narrow scale, depending on the composition and concentration of salts in the soil. However, different sets of soil characteristics influence the distribution pattern of species and communities (Castañeda et al. 2013; Stevanovic et al. 2019).

We find a range of halophyte species suitable to grow in different saline regions globally. These are: coastal saline soil, soils of mangrove forests, wetlands, marshy lands, arid and semiarid regions, and agricultural fields (Kumar et al. 2019). Such plants do represent an ideal model to understand complex physiological and genetic mechanisms of salinity stress tolerance. An effective improvement of saline soils is possible by growing halophytes, as these are well adjusted to the saline environment because of their diversified adaptation mechanisms such as succulence, osmotic adjustment, ion compartmentalization- transport - uptake, salt inclusion or excretion, antioxidant systems and maintenance of redox status (Lokhande and Suprasanna 2012; Kumar et al. 2019).

Lately, halophytes have been identified, classified and studied by many investigators notable among these being von Sengbusch (2003), Devi et al. (2019), and Kumar et al. (2019). These researchers have reported that halophytes can be classified on their habitat features as "obligate halophytes," "optional halophytes," and "habitat indifferent halophytes." **Obligate halophytes** are specific to saline habitats and thrive well only in saline environments, e.g., Chenopodiaceae family (Devi et al. 2019; Kumar et al. 2019). **Facultative halophytes** can grow on salt-affected soils, except that their threshold performance is possible only under minimum saline or no-saline conditions, e.g., majorly Poaceae and Cyperaceae families (Devi et al. 2019; Kumar et al. 2019). **Habitat-indifferent halophytes** are the ones not necessarily native to saline habitats but are capable of living on such soils. As exotics of the saline habitats they compete with plants sensitive to salt and under extreme conditions are capable of salt tolerance (Devi et al. 2019; Kumar et al. 2019).

Another ecological grouping is the classification of plants based on eco-physiological features in saline habitats (Devi et al. 2019):

1. *Salt-requiring halophytes:*

   (a) **Obligatory**: These taxa depend on salts for their growth and survival like *Salicornia* sp.
   (b) **Preferential halophytes**: The growth and development are improved in such taxa in the presence of salts like *Aster*, *Nitaria,* and *Suaeda* species.
   (c) **Salt-enduring**: Enduring a high protoplasmic salt content such as *Suaeda monoica*.

2. *Salt-resistant halophytes:*

   (a) **Salt-excluding**: These accumulate salts in specific hairs; and secrete salts from their shoots like *Atriplex*, *Aeluropus*, *Limonium*, and *Tamarix* species.
   (b) **Salt-evading**: Keeping away from the salt uptake; avoid salt transport into their leaves like *Rhizophora* species and *Prosopis foxia*.

In addition to these classification systems the plants distributed on saline habitats have also been classified into three categories by Rajpurohit (1980):

1. **True halophytes** show optimum growth only on saline soil (above 0.5% NaCl level).
2. **Facultative halophytes** very well growing on saline (at 0.5% NaCl level) soils vis-a-vis nonsaline soils, behaving as halophytes or glycophytes, respectively.
3. **Glycophytes** that is the plants of non-saline habitat showing their optimal growth here.

Breckle (1995) has classified halophytes into three categories depending on the mode of ion accumulation and transport:

1. **Recretohalophytes**: (a) Exo-recretohalophytes - exluding the salt through leaf salt glands (b) Endo-recretohalophytes - excluding the salt through leaf vesicles (Breckle 1995; Li et al. 2019).
2. **Euhalophytes**: (a) Leaf succulent euhalophytes, these accumulate the salt ions in vacuoles of succulent leaves. The "Stem succulent euhalophytes" in the central column (Breckle 1995; Li et al. 2019).
3. **Pseudo-halophytes**: Accumulating salt ions in the vacuoles and root xylem parenchyma (Breckle 1995; Li et al. 2019).

All three grow on saline-alkali land in China (Li et al. 2019).

The halophytes have been classified by several authors based on ecological types as "**Xerohalophytes, Psammophytes, and Hydrohalophytes**" (Khan 2003; Altay and Ozturk 2012; Guvensen et al. 2006; Ozturk et al. 2006, 2008, 2014a, 2016; Bueno and Cordovilla 2021). Xerohalophytes includes plants which can tolerate a certain amount of drought and salt in zones where they live. The salts are retained in the soil because of the great evaporation of water, as such they are also subjected to varying salinity conditions, depending on the type of soil and rainfall year round

(Bueno and Cordovilla 2021). The sandy habitats are the places occupied by psammophytes because they are adapted to living on such habitats, usually having mobility (dunes). They too are saline when influenced by the sea (Bueno and Cordovilla 2021). Hydrohalophytes are capable of surviving in areas subject to flooding and therefore tolerate fluctuations in salinity levels in their habitats (Bueno and Cordovilla 2021).

Halophytes belonging to different ecological groups are presented in Table 1.3.

Table 1.3 Some of the halophytes belonging to different ecological groupings

Ecological groupings – Halophytes samples
Obligate halophytes
Aeluropus lagopoides, Arthrocnemum macrostachyum, Avicennia marina, Caroxylon spp., *Cyperus aucheri, Halocnemum strobilaceum, Halopeplis perfoliata, Juncus rigidus, Limonium* spp., *Odyssea mucronata, Salicornia perennans, Seidlitzia rosmarinus, Sporobolus spicatus, S. consimilis, Suaeda* spp., *Tamarix* spp., *Tetraena* spp., and *Urochondra setulosa*
Facultative halophytes
Anabasis setifera, Salsola drummondii, Suaeda vermiculata, S. aegyptiaca, and *Tetraena qatarense*
Recretohalophytes (exo-recretohalophytes)
Acanthus ebracteatus, Aegiceras corniculatum, Aeluropus sp., *Castilleja pallida, Cenchrus* sp., *Crypsis* sp., *Frankenia pulverulenta, Glaux maritima, Ipomoea polymorpha, I. pes-tigridis, Limonium bicolor, L. aureum, Reaumuria* sp., *Spartina* sp., *Sporobolus* sp., and *Tamarix* spp.
Recretohalophytes (endo-recretohalophytes)
Atriplex, Chenopodium, and *Salsola* species
Euhalophytes (leaf succulent euhalophytes)
Suaeda salsa, S. glauca, Salsola scopparia, and *S. junatoxxii*
Euhalophytes (stem succulent euhalophytes)
Halostachys belonggeriana, Halocnemum strobilaceum, Kalidium schrenkianum, K. cuspidatum, and *Salicornia europaea*
Pseudo-halophytes
Lotus siliquosus, Phragmites communis, and *Trifolium fragiferum*
Xerohalophytes
Atriplex halimus, Chenopodium quinoa, Tamarix ramosissima, and *Zygophyllum xanthoxylum*
Psammophytes
Cakile maritima, Distichlis spicata, Ficinia nodosa, Halopyrum mucronatum, Haloxylon ammodendron, Sporobolus virginicus, and *Urochondra setulosa*
Hydrohalophytes
Arthrocnemum macrostachyum, Juncus acutus, Phragmites australis, Salicornia europaea, and *Sarcocornia fruticosa*

Modified from Khan (2003), Altay and Ozturk (2012), Ozturk et al. (2014a, 2016), Ghazanfar et al. (2019), Li et al. (2019), Bueno and Cordovilla (2021)

1.5 Panoramic Views of Saline Habitats

Halophytes characterize complex and heterogeneous ecological taxa containing a variety of plants with different adaptive characteristics and taxonomical and habitat diversity (Grigore and Toma 2017). These include a cluster of taxa which complete their life cycle and growth in a habitat with great salt content (Grigore and Toma 2010). Salt tolerant taxa can reproduce in the salt concentration of 200 Mm NaCl or more (Flowers and Colmer 2008). Halophytic taxa are able to grow in varied environmental situations dependent on the tolerance and response ability to several salts in the substrate. Halophytes can be categorized in many ways such as inland, coastal, and near coastal plants dependent on their habitats, i.e., salt marsh or desert (Song et al. 2008).

Some specific habitat types where halophytes are naturally distributed are summarized in Table 1.4.

Table 1.4 Some important halophytic habitat types

Coastal habitats
Coastal dunes
Coastal littoral plain areas
Coastal saltmarshes / littoral salt marsh
Coastal stable dune grasslands (gray dunes)
Humid and semi-humid areas of seashore
Low coastal dunes which receive salt spray
Marine mud shores
Sandy coastal plain
Sandy shores
Saline sand formations along beaches
Seagrass beds on littoral sediments
Inland habitats
Alpine steppe
Brackish lakes, ponds and pools
Clayey-salty, relatively wet areas
Continental inland salt steppes
Dry grasslands
Inland salt marshes / terrestrial salt marshes
Inland sandy saline plains
Karstic limestone plateau areas
Permanent inland saline areas
Saline reedbeds
Saline, silt plains
Sandy areas around lakes
Sandy plains
Semi-humid and semiarid meadow steppe

(continued)

Table 1.4 (continued)

Semi-humid and semiarid tillage steppe
Saline steppe
Salt steppes
Desert and sabkha habitats
Arid desert
Arid desert steppe
Dune valleys
Dunes and flat sandy areas
Extreme arid region desert
Oases
Open salt desert
Open sand desert
Salt desert
Sandy desert
Sandy-salty depression areas
Semi-desert areas
Semiarid semidesert steppe
Shallow sands
Stabilized dunes

Modified from eHALOPH (2019), EUNIS (2019), Kapler (2019), Li et al. (2019), Obón et al. (2021)

The halophytes are using amazing diversity of cell and tissue mechanisms an based on their ecophysiological processes which shed new light on adaptations to extreme environments (Kapler 2019). In the case of phylogenetic origins of the present diversity in halophytes a better understanding is required which can be highly critical for obtaining novel organisms for land reclamation, phytoremediation and salt pan-based agriculture (Flowers et al. 2010). As per Bosiacka et al. (2016), we are indebted to the genetic investigations as well as thorough morphological studies on the saline habitat plants because the researchers have started to discover the hidden species diversity of the halophilous and subhalophilous coastal and inland grasslands. The coastal salt marshes and meadows are threatened by reclamation for agriculture, destruction for urbanization and tourisim facilities as well as dumping of organic and inorganic wastes together with sand removal. These destructive practices should be controlled strictly to save these ecosystems (Ozturk et al. 1995, 2006; Guvensen and Ozturk 2003; Guvensen et al. 2006; Tug et al. 2011; Altay and Ozturk 2012).

Panoramic views of saline habitats in some countries of the Asian continent are presented in Figs. 1.1, 1.2, 1.3, 1.4, 1.5, 1.6, 1.7, 1.8, 1.9, and 1.10.

Fig. 1.1 Halophytic flora of Salt Range of Soon Sakesar, Khushab, Punjab, Pakistan

Fig. 1.2 Jashak Salt Dome, Iran

Fig. 1.3 PMDC Warcha Salt Mines, Khushab, Punjab, Pakistan

Fig. 1.4 Panoramic view of Asia's largest Khewra Salt Range, Jhelum, Punjab, Pakistan

Fig. 1.5 Inland salt marshes and salt steppes in Tuz Lake, Turkiye

Fig. 1.6 Mangrove forests on the coast of Malaysia

Fig. 1.7 Coastal salt marshes of the Aegean region of Turkiye

Fig. 1.8 Sandy shores of the Agean region of Turkiye

Fig. 1.9 Humid and semi-humid areas of the seashore in the United Arab Emirates

Fig. 1.10 Wet solonchaks of Bukhara oasis in Uzbekistan

References

Abbas J (2002) Plant communities bordering the sabkhat of Bahrain Island. In: Barth H-J, Böer B (eds) Sabkha ecosystems, The Arabian Peninsula and adjacent countries, vol 1. Kluwer Academic, Dordrecht, pp 51–62

Abbas JA, El-Oqlah AA (1992) Distribution and communities of halophytic plants in Bahrain. J Arid Environ 22:205–218

Abdel-Razik MS (1991) Population structure and ecological performance of the mangrove Avicennia marina (Forssk.) Vierh. On the Arbian Gulf coast of Qatar. J Arid Environ 20:331–338

Abdel-Razik MS, Ismail AM (1990) Vegetation composition of a maritime salt marsh in Qatar in relation to edaphic factors. J Veg Sci 1:85–88

Akcin AT, Akcin A, Yalcin E (2017) Anatomical changes induced by salinity stress in *Salicornia freitagii* (Amaranthaceae). Rev Bras Bot 40:1013–1018

Akhani H (2004) Halophytic vegetation of Iran: towards a syntaxonomical classification. Annales Botanici Nuova Ser 4:65–82

Akhani H (2006) Biodiversity of halophytic and sabkha ecosystems in Iran. In: Khan MA et al (eds) Sabkha ecosystems, West and Central Asia, vol II. Springer, Dordrecht, pp 71–88

Al Khulaidi AA (2013) Flora of Yemen. The Sustainable Natural Resource Management Project (SNRMP II) EPA and UNDP, Republic of Yemen

Al Khulaidi AA, Miller AG, Furley P (2010) Environmental and human determinates of vegetation distribution: in the Hadhramaut region. LAP Lambert Academic Publishing, Saarbrücken, p 420

Al-Gifri AN, Gabali SA (2002) The coastal sabkhat of Yemen. In: Barth H-J, Böer B (eds) Sabkha ecosystems, The Arabian Peninsula and adjacent countries, vol 1. Kluwer Academic, Dordrecht, pp 141–146

Al-Taisan WA, Gabr DG (2017) Comparative morphological and anatomical characters of *Cakile arabica* from different habitat in eastern region of Saudi Arabia. Saudi J Biol Sci 24:226–233

Altay V (2012) Mustafa Kemal Üniversitesi Tayfur Ata Sökmen Kampüsü (Hatay)'nün süs bitkileri. Karadeniz Fen Bilimleri Dergisi 2(6):11–26

Altay V (2019) Ecology of *Pinus sylvestris* L. Forests – a case study from Istanbul (Turkey). Pak J Bot 51(5):1711–1718

Altay V, Çelik O (2011) Antakya semt pazarlarındaki bazı doğal bitkilerin etnobotanik yönden araştırılması. Biyoloji Bilimleri Araştırma Dergisi 4(2):137–139

Altay V, Karahan F (2012) Tayfur Sökmen Kampüsü (Antakya-Hatay) ve çevresinde bulunan bitkiler üzerine etnobotanik bir araştırma. Karadeniz Fen Bilimleri Dergisi 2(7):13–28

Altay V, Karahan F (2017a) Anadolu geleneksel tıbbında güneş çarpması ve güneş yanığı tedavisinde kullanılan tıbbi bitkiler. Erzincan Univers J Sci Technol 10(1):124–137

Altay V, Karahan F (2017b) Ruderal vejetasyon üzerine bir ön çalışma. Kilis 7 Aralık Üniversitesi Fen ve Mühendislik Dergisi 1(2):68–77

Altay V, Ozturk M (2012) Land degradation and halophytic plant diversity of Milleyha wetland ecosystem (Samandag-Hatay) Turkey. Pak J Bot 44:37–50

Altay V, Ozturk M (2021) The genera *Salsola* and *Suaeda* (Amaranthaceae) and their value as fodder. In: Grigore M-N (ed) Handbook of halophytes. Springer, Cham. https://doi.org/10.1007/978-3-030-17854-3_97-1

Altay V, Ozyıgıt II, Yarcı C (2010a) Urban flora and ecological characteristics of the Kartal District (Istanbul): a contribution to urban ecology in Turkey. Sci Res Essays 5(2):183–200

Altay V, Ozyıgıt II, Yarcı C (2010b) Urban ecological characteristics and vascular wall flora on the Anatolian side of Istanbul, Turkey. Maejo Int J Sci Technol 4(3):483–495

Altay V, Çelik O, Kayıkçı S (2011) Hatay'ın vasküler duvar florası. Ot Sistematik Botanik Dergisi 18(2):131–144

Altay V, Ozyıgıt II, Yarcı C (2012a) Plant communities in urban habitats of Istanbul-Turkey. Pak J Bot 44:177–186

Altay V, Serın M, Yarcı C, Severoglu Z (2012b) Phytoecological and phytosociological ınvestigations of the vegetation of Gölcuk (Kocaeli/Turkey). Ekoloji 21(84):74–89

Altay V, Ozyığıt II, Keskin M et al (2013) An ecological study of endemic plant *Polygonum istanbulicum* Keskin. Pak J Bot 45:455–459
Altay V, Keskin M, Karahan F (2015a) An assessment of the plant biodiversity of Mustafa Kemal University Tayfur Sokmen Campus (Hatay-Turkey) for the view of human health. Int J Sci Technol Res 1(2):83–103
Altay V, Ozyığıt II, Osma E et al (2015b) Enviromental relationships of the vascular flora alongside the railway tracks between Haydarpasa and Gebze (Istanbul-Kocaeli/Turkey). J Environ Biol 36:153–162
Altay V, Karahan F, Sarcan YB, Ilçım A (2015c) An ethnobotanical research on wild plants sold in Kırıkhan district (Hatay/Turkey) herbalists and local markets. Biol Divers Conserv 8(2):81–91
Altay V, Karahan F, Ozturk M et al (2016a) Molecular and ecological investigations on the wild populations of *Glycyrrhiza* L. taxa distributed in the East Mediterranean Area of Turkey. J Plant Res 129(6):1021–1032
Altay V, Gulyanar Ş, Ozyıgıt II (2016b) Autecology of *Cephalaria taurica* Szabó, a narrow endemic from Turkey: plant-soil interactions. IOSR J Environ Sci Toxicol Food Technol 10(9):90–94
Altay V, Daloğlu MY, Ozturk M (2017) Edaphic relations of *Cirsium cassium* Davis & Parris (Asteracee), a local endemic from Hatay (Turkey). Anatol J Bot 1(2):41–44
Altay V, Karahan P, Karahan F, Ozturk M (2018) Pollen analysis of honeys from Hatay/Turkey. Biol Divers Conserv 11(3):209–222
Altay V, Silc U, Yarcı C et al (2020) Urban vegetation of the Anatolian side of Istanbul. Phytocoenologia 50(2):101–121
Aronson J (1989) HALOPH; salt tolerant plants for the world – a computerized global data base of halophytes with emphasis on their economic uses. University of Arizona Press, Tucson
Ayub M, Ashraf MY, Kausar A, Saleem S, Anwar S et al (2021) Growth and physio-biochemical responses of maize (*Zea mays* L.) to drought and heat stresses. Plant Biosyst 155(3):535–542
Babikir AA (1984) Vegetation and environment on the coastal sand, dunes and playas of Khor El-Odaid Area, Qatar. Geo J 9:377–385
Babikir AA, Kürschner H (1992) Vegetational patterns within a coastal saline of NE-Qatar. Arab Gulf J Sci Res 10:61–75
Bahadur S, Ahmad M, Mir S et al (2018) Identification of monocot flora using pollen features through scanning electron microscopy. Microsc Res Tech 81(6):599–613
Batanouny KH (1981) Ecology and flora of Qatar. Qatar University Press, Doha
Batanouny KH, Turki AA (1983) Vegetation of South-Western Qatar. Arab Gulf J Sci Res 1:5–19
Batanouny KH, Hassan AH, Fahmy GM (1992) Eco-physiological studies on halophytes in arid and semi-arid zones. II. Eco-physiology of *Limonium delicatulum* (Gır.) Kt. Flora 186:105–116
Böer B (1994) Status, environmental factors and recovery of the intertidal and terrestrial vegetation between Ras as-Zaur and Abu Ali Island after the Gulf war oil spill. In: Establishment of a marine habitat and wildlife sanctuary for the Gulf region. Final report for phase II. CEC/NCWCD, Frankfurt/Jubail, pp 229–253
Böer B (1996) Trial planting of mangroves (Avicennia marina) and salt marsh plants (Salicornia europaea) in oil-impacted soil in Jubail area, Saudi Arabia. In: Krupp F, Abuzinada AA, Nader JA (eds) A marine wildlife sanctuary for the Arabian Gulf. National Commission for Wildlife Conservation and Development, Riyadh, pp 186–192
Böer B (2002) The coastal and sabkha flora of the United Arab Emirates. Short communication In: Barth HJ, Böer B (eds) 2002: Sabkah ecosystems vol. I: the Arabian peninsula and adjacent countries. Afghanistan, Pakistan, Iran, Jordan, Kuwait, Saudi Arabia, Bahrain, Qatar, United Arab Emirates, Oman, Yemen, Egypt, Sudan, Eritrea, Ethiopia, Djibouti, Somalia. Tasks for vegetation science 36. Kluwer Academic Publishers, pp 303–309
Böer B, Al Hajiri S (2002) The coastal and sabkha flora of Qatar: an introduction. In: Barth H-J, Böer B (eds) Sabkha ecosystems, The Arabian Peninsula and adjacent countries, vol 1. Kluwer Academic, Dordrecht, pp 63–70
Böer B, Saenger P (2006) The biogeography of the coastal vegetation of the Abu Dhabi gulf coast. In: Khan A, Böer B, Kust GS, Barth H-J (eds) Sabkah ecosystems vol. II: West and Central Asia, Tasks for vegetation science, vol 42. Springer, Heidelberg, pp 31–36

Böer B, Warnken J (1992) Qualitative analysis of the coastal and inland vegetation of the Dawkat ad-Dafi and Dawkat al-Mussalamiya region. In: Establishment of a marine habitat and wildlife sanctuary for the Gulf region. Final report for phase I. CEC/NCWCD, Frankfurt/Jubail, pp 81–101

Borhidi A, Kevey B, Lendvai G (2012) Plant communities of Hungary. Akadémiai Kiadó, Budapest

Bosiacka B, Kull T, Więcław H, Marciniuk P, Podlasiński M (2016) Habitat requirements of marsh dandelions (*Taraxacum*) in Polish and Estonian coastal grasslands. Pol J Ecol 64(2):213–230

Breckle SW (1995) How do halophytes overcome salinity? In: Khan MA, Ungar IA (eds) Biology of salt tolerant plants. Department of Botany, University of Karachi, Karachi, pp 199–213

Breckle SW (2016) Halophytes and saline vegetation of Afghanistan, a potential rich source for people. In: Khan MA et al (eds) Halophytes for food security in dry lands. Academic Press, pp 49–66

Breckle SW (2021) An ecological overview of halophytes from the Aralkum Area. In: Grigore MN (ed) Handbook of halophytes: from molecules to ecosystems towards biosaline agriculture. Springer, Cham, pp 393–449

Breckle S-W, Agachanjanz O, Wucherer W (1998) Der Aralsee: Geoökologische Probleme. Naturwissenschaftliche Rundschau 9:347–355

Breckle S-W, Wucherer W, Agachanjanz O, Geldyev B (2001) The Aral Sea crisis region. In: Breckle S-W, Veste M, Wucherer W (eds) Sustainable land use in deserts. Springer, Berlin, pp 27–37

Brown G, Böer B, Sakkir S (2008) The coastal vegetation of the western and southern Gulf – characterisation and conservation aspects. In: Abuzinada AH, Barth H-J, Krupp F et al (eds) Protecting the Gulf's marine ecosystems from pollution. Birkhauser Verlag, Basel, pp 23–44

Bueno M, Cordovilla MP (2021) Ecophysiology and uses of halophytes in diverse habitats. In: Grigore M-N (ed) Handbook of halophytes: from molecules to ecosystems towards biosaline agriculture, vol II. Springer, Cham, pp 1613–1636

Caparrós PG, Ozturk M, Gul A, Batool TS, Pirasteh-Anosheh H et al (2022) Halophytes have potential as heavy metal phytoremediators: a comprehensive review. Environ Exp Bot 193:104666

Castañeda C, Herrero J, Conesa JA (2013) Distribution, morphology and habitats of saline wetlands: a case study from Monegros, Spain. Geol Acta 11:371–388

Chaudhary SA (1998) Flora of the Kingdom of Saudi Arabia 1: 14–15. Ministry of Agriculture and Water. National Agriculture and Water Research Centre, Riyadh

Chia-Chi H, Tsen-Li C (1993) Halophytes in China: floristic distribution and vegetation type. In: Lieth H, Al Masoom AA (eds) Towards the rational use of high salinity tolerant plants, Tasks for vegetation science, vol 27. Springer, Dordrecht. https://doi.org/10.1007/978-94-011-1858-3_3

Clarke AE, Andreson RL, Stone BA (1979) Form and function of arabinogalactans and arabinogalactan-proteins. Phytochemistry 18:521–540

Corbishley J, Pearce D (2007) Growing trees on salt-effected land. ACIAR impact assessment series report no. 51 ACIAR

Dagar JC (1995) Characteristics of halophytic vegetation in India. In: Khan MA, Ungar IA (eds) Biology of salt tolerant plants. Department of Botany, University of Karachi, Karachi, pp 255–276

Dajic Z (2006) Salt stress – salinity and tolerance mechanisms in plants. In: Madhava Rao KV, Raghavendra AS, Reddy KJ (eds) Physiology and molecular biology of stress tolerance in plants. Springer, Dordrecht, pp 41–99

Deil U, Müller-Hohenstein K (1996) An outline of the vegetation of Dubai (UAE). Verhandlungen der Gesellschaft für Ökologie 25:77–95

Devi S, Kumar A, Arya SS, Kumari A, Kumar N et al (2019) Economic utilization and potential of halophytes. In: Hasanuzzaman M et al (eds) Ecophysiology, abiotic stress responses and utilization of halophytes. Springer, Singapore, pp 195–220

Dickison WC (2000) Integrative plant anatomy. Harcourt Academic Press, San Diego

Ding F, Yang JC, Yuan F, Wang BS (2010) Progress in mechanism of salt excretion in recretohalopytes. Front Biol 5:164–170

Egamberdieva D, Ozturk M (eds) (2018) Vegetation of central Asia and environs. Springer, Cham
Ehaloph (2019) Halophytes database – University of Sussex. https://www.sussex.ac.uk/affiliates/halophytes
El-Demerdash MA, Hegazy AK, Zilay MA (1995) Vegetation-soil relationship in Tihamah coastal plains of Jazan region, Saudi Arabia. J Arid Environ 30:161–174
El-Sheikh AM, Youssef MM (1981) Halophytic and xerophytic vegetation near al Kharj springs. Bull Fac Sci King Saud Univ 12:5–21
El-Sheikh MA, Mahmoud A, El-Tom M (1985) Ecology of the inland salt marsh vegetation at Al-Shiggat in Al-Qassim district, Saudi Arabia. Arab Gulf J Sci Res 3:165–182
El-Shourbagy MN, Al-Eidaros OH, Al-Zahrani HS (1987) Distribution of Halopeplis perfoliata (Forssk.) Bge. ex Schweinf. In the Red Sea coastal salt marshes: phytosociological relations and response to soil. J Coast Res 3:179–187
Eroglu HK, Ozyıgıt II, Altay V, Yarcı C (2014) Autecological characteristics of *Centaurea hermannii* F. Herm. an endemic species from Turkey. Bulgarian J Agr Sci 20:183–187
Eskin B, Altay V, Ozyıgıt II, Serın M (2012) Urban vascular flora and ecological characteristics of the Pendik District (Istanbul-Turkey). Afr J Agric Res 7(4):629–646
Eskin B, Ozyıgıt II, Dogan I et al (2013) Germination physiology and autecology of Centaurea kilaea Boiss. from Turkey. Sains Malaysiana 42(10):1473–1482
EUNIS (2019) European nature information system. https://eunis.eea.europa.eu/index.jsp. Last accessed 19 Nov 2019
Fahn A (1988) Secretory tissues in vascular plants. New Phytol 108:229–257
Flowers TJ, Colmer TD (2008) Salinity tolerance in halophytes. New Phytol 179:945–963
Flowers TJ, Colmer TD (2015) Plant salt tolerance: adaptations in halophytes. Ann Bot 115(3):327–331
Flowers TJ, Galal HK, Bromham L (2010) Evolution of halophytes: multiple origins of salt tolerance in land plants. Funct Plant Biol 37(7):604–612
Füzy A, Bíró B, Tóth T (2010) Effect of saline soil parameterson endomycorrhizal colonisation of dominant halophytes in four Hungarian sites. Span J Agric Res 8:144–148
Ghanem ME, Han RM, Classen B et al (2010) Mucilage and polysaccharides in the halophyte plant species *Kosteletzkya virginica*: localization and composition in relation to salt stress. J Plant Physiol 167:382–392
Ghazanfar SA (1992) Quantitative and biogeographic analysis of the flora of the Sultanate of Oman. Glob Ecol Biogeogr Lett 2:189–195
Ghazanfar SA (1993) Vegetation of the khawrs and fresh water Springs of Dhofar. Part E. In: Khawrs and Springs of the Dhofar Governorate. Survey and monitoring studies. Unpublished report, Planning Committee for Development and Environment in the Governorate of Dhofar, Oman
Ghazanfar SA (1995) Coastal sabkhas: an analysis of the vegetation of Barr al Hikman. In: Khan MA, Ungar IA (eds) The biology of salt tolerant plants. Department of Botany, University of Karachi, Karachi, pp 277–283
Ghazanfar SA (1998) Water vegetation. In: Ghazanfar SA, Fisher M (eds) Vegetation of the Arabian Peninsula. Kluwer Academic Press, Dordrecht, pp 229–240
Ghazanfar SA (1999) Coastal vegetation of Oman. Estuar Coast Shelf Sci 49:21–27
Ghazanfar SA (2002) The sabkha vegetation of Oman. In: Barth H-J, Böer B (eds) Sabkha ecosystems, The Arabian Peninsula and adjacent countries, vol 1. Kluwer Academic, Dordrecht, pp 99–108
Ghazanfar SA (2006a) Sabkhat regions of Iraq. In: Khan MA et al (eds) Sabkha ecosystems, II: West and Central Asia. Springer, Dordrecht, pp 211–217
Ghazanfar SA (2006b) Saline and alkaline vegetation of NE Africa and the Arabian Peninsula: An overview. In: Ozturk M, Waisel Y, Khan MA, Gork G (eds) Biosaline agriculture and salinity tolerance in plants. Birkhaeuser Publishing, Basel, pp 101–108
Ghazanfar SA, Rappenhöner D (1994) Vegetation and flora of the islands of Masirah and Shaghaf, Sultanate of Oman. Arab Gulf J Sci Res 12(3):509–524

Ghazanfar SA, Altundag E, Yaprak AE, Osborne J, Tug GN, Vural M (2014) Halophytes of Southwest Asia. In: Khan MA et al (eds) Sabkha ecosystems: volume IV – cash crop halophyte and biodiversity conservation, Tasks for vegetation science, vol 47. Springer, Dordrecht, pp 105–133

Ghazanfar SA, Böer B, Khulaidi AWA et al (2019) Plants of sabkha ecosystems of the Arabian Peninsula. In: Gul B et al (eds) Sabkha Ecosystems. Springer, pp 55–80

Grigore MN (ed) (2021) Handbook of halophytes – from molecules to ecosystems towards biosaline agriculture. Springer, Cham. https://doi.org/10.1007/978-3-030-57635-6

Grigore MN, Toma C (2010) Halofitele. Aspecte De Anatomie Ecologica. Universitatii Alexandru Ioan Cuza, Iasi, Romania

Grigore MN, Toma C (2017) Anatomical adaptations of halophytes. A review of classic literature and recent findings. Springer, Cham

Grigore MN, Ivanescu L, Toma C (2014) Halophytes: an integrative anatomical study. Springer, Cham

Guvensen A, Ozturk M (2003) Halophytic plant diversity of South Aegean coastal zone in Turkey. Pak J Bot 35(5):853–864

Guvensen A, Gork G, Ozturk M (2006) An overview of the halophytes in Turkey. In: Khan MA, Boer B, Kust GS, Barth HJ (eds) Sabhka ecosystems II: west and Central Asia. Springer, Dordrecht, pp 9–30

Hajibagheri MA, Hall JL, Flowers TJ (1984) Stereological analysis of leaf cells of the halophyte *Suaeda maritima* (L.) Dum. J Exp Bot 35:1547–1557

Halwagy R (1986) On the ecology and vegetation of Kuwait. In: Kürschner H (ed) Contributions to the vegetation of Southwest Asia. Beihefte zum Tübinger Atlas des Vorderen Orients, Reihe A (Naturwissenschaften), vol 24. Dr Ludwig Reichert Verlag, Wiesbaden, pp 81–109

Halwagy R, Halwagy M (1977) Ecological studies on the desert of Kuwait. III. The vegetation of the coastal salt marshes. J Univ Kuwait (Sci) 4:33–73

Halwagy R, Moustafa AF, Kamal S (1982) On the ecology of the desert vegetation in Kuwait. J Arid Environ 5:95–107

Hameed M, Ashraf M, Ahmad MSA, Naz N (2010a) Structural and functional adaptations in plants for salinity tolerance. In: Asharaf M, Ozturk M, Ahmad MSA (eds) Plant adaptation and phytoremediation. Springer, Berlin, pp 151–173

Hameed M, Ashraf M, Naz N, Al-Qurainy F (2010b) Anatomical adaptations of *Cynodon dactylon* (L.) Pers., from the salt range Pakistan, to salinity stress. I. Root and stem anatomy. Pak J Bot 42:279–289

Hameed A, Qadri TN, Mahmooduzzafar TOS et al (2017) Physio-biochemical and nutritional responses of Abelmoschus esculentus (L.) Moench (Okra) under mercury contamination. Fresenius Environ Bull 26:5814–5823

Haq S, Bharose R, Ozturk M, Altay V, Bhattı AA, Dervash MA, Hakeem KR (2021) Impact of treated sewage water on nutrient status of alfisols and vegetable crops. Notulae Botanicae Horti Agrobotanici Cluj-Napoca 49(2):12255–12255

Harrison S, Yu G, Takahara H, Prentice I (2001) Diversity of temperate plants in east Asia. Nature 413(6852):129–130

Hasegawa PM, Bressan RA, Zhu J-K, Bohnert HJ (2000) Plant cellular and molecular responses to high salinity. Annu Rev Plant Physiol Plant Mol Biol 51:463–499

Heathcote JA, King S (1998) Umm as Samim, Oman: a sabkha with evidence for climatic change. In: Alsharhan AS, Glennie KW, Whittle GL, Kendall CGSC (eds) Quaternary deserts and climatic change. Balkema, Rotterdam

Hocaoglu-Ozyigit A, Ucar B, Altay V, Ozyigit II (2022) Genetic diversity and phylogenetic analyses of Turkish cotton (Gossypium hirsutum L.) lines using ISSR markers and chloroplast trnL-F regions. J Nat Fibers 19(5):1837–1850

Imanberdieva N, Severoğlu Z, Kurmanbekova G et al (2018a) Chapter 16: plant diversity of Ala-Archa Nature Park in Kyrgyzstan with emphasis on its economic potential. In: Egamberdieva D, Ozturk M (eds) Vegetation of Central Asia and environs. Springer, pp 365–381

Imanberdieva N, Imankul B, Severoğlu Z et al (2018b) Chapter 15: potential impacts of climate change on plant diversity of Sary-Chelek Biosphere Reserve in Kyrgyzstan. In: Egamberdieva D, Ozturk M (eds) Vegetation of Central Asia and environs. Springer, Cham, pp 349–364
Iqbal A, Razzaq A, Hadı F et al (2018) Assesment of genetic diversity among hybrid pea lines (Pisum sativum L.) as revealed by random amplified polymorphic DNA (RAPD) markers. Fresenius Environ Bull 27(10):6447–6453
Jaiswar S, Kazi MA (2016) Composition and localization of mucilage under short term NaCl stress in *Salicornia brachiata* Roxb. and *Suaeda maritima* (L.) Dumort. Indian J Nat Prod Resour 7:162–168
Kafi M, Kamili AN, Husaini AM et al (2018) Chapter 4: an expensive spice – saffron (*Crocus sativus* L.) – case study from Kashmir, Iran and Turkey. In: Ozturk M et al (eds) Global perspectives on underutilized crops. Springer, pp 109–149
Kapler A (2019) Habitats of halophytes. In: Hasanuzzaman M, Shabala S, Fujita M (eds) Halophytes and climate changes: adaptive mechanism and potential uses. CAB International, Boston
Karahan F, Çelik O, Kayıkçı S, Altay V (2012a) Eski Antakya evleri (Antakya-Hatay) duvarlarında yayılış gösteren vasküler bitkiler. Biyoloji Bilimleri Araştırma Dergisi 5(2):131–134
Karahan F, Çelik O, Kayıkçı S, Altay V (2012b) Antakya (Hatay)'nın ayakaltı bitkileri. Biyoloji Bilimleri Araştırma Dergisi 5(2):135–137
Karahan F, Altay V, Keskin M (2015) An ethnobotanical study on woody plants benefits from handicrafts in Antakya District (Hatay-Turkey). Int J Sci Technol Res 1(1):1–18
Kelemen A, Török P, Valkó O, Miglécz T, Tóthmérész B (2013) Mechanisms shaping plant biomass and species richness: plant strategies and litter effect in alkali and loess grasslands. J Veg Sci 24:1195–1203
Khan MA (2003) An ecological overview of halophytes from Pakistan. In: Lieth H (ed) Cash crop halophytes: recent studies. Kluwer Academic Publishers, London, pp 167–187
Khan MA, Qaiser M (2006) Halophytes of Pakistan: characteristics, distribution and potential economic usages. In: Sabkha ecosystems, vol 42. Springer, Dordrecht, pp 129–153
Khan R, Abidin U, Zain S et al (2018) Palyno-morphological characteristics of gymnosperm flora of Pakistan and its taxonomic implications with LM and SEM methods. Microsc Res Tech 81(1):74–87
Kokab S, Ahmad S (2010) Characterizing salt tolerant plants using ecosystem and economic utilization potentials for Pakistan. Manag Nat Resourr Sustain Future Agric 2(12):1–20
Koyro HW, Geißler N, Hussin S et al (2008) Strategies of halophytes to survive in a salty environment. In: Khan NA, Singh S (eds) Abiotic stress and plant responses. I.K. International Publishing House, New Delhi, pp 83–104
Krishnamurthy P, Jyothi-Prakash PA, Qin L et al (2014) Role of root hydrophobic barriers in salt exclusion of a mangrove plant *Avicennia officinalis*. Plant Cell Environ 37:1656–1671
Kumar N, Shubham L, Ashwani K, Pratima K, Mann A et al (2019) Antioxidant defence in halophytes under high salinity. In: Hasanuzzaman M, Shabala S, Fujita M (eds) Halophytes and climate change: adaptive mechanisms and potential uses. CAB International, Wallingford, pp 196–208
Kürschner H (1986) A study of the vegetation of the Qurm Nature Reserve, Muscat area, Oman. Arab Gulf J Sci Res 4:23–52
Kürschner H, Al-Gifri AN, Al-Subai MY, Rowaished AK (1998) Vegetation pattern within coastal salines of southern Yemen. Feddes Repert
L'vovich MI (1974) Global water resources and their future. Mysl', Moscow
Letchamo W, Ozturk M, Altay V et al (2018) Chapter 2: A alternative potential natural genetic resource: Seabuckthorn [*Elaeagnus rhamnoides* (syn.: *Hippophae rhamnoides*)]. In: Ozturk M et al (eds) Global perspectives on underutilized crops. Springer, pp 25–82
Li G, Zhao P, Shao W (2019) Cash crop halophytes of China. In: Gul B et al (eds) Sabkha ecosystems, Tasks for vegetation science, vol VI. Springer, Cham, pp 497–504

Liang L, Liu W, Sun Y et al (2017) Phytoremediation of heavy metal contaminated saline soils using halophytes:current progress and future perspectives. Environ Rev 25:269–281
Lokhande VH, Suprasanna P (2012) Prospects of halophytes in understanding and managing abiotic stress tolerance. In: Ahmad P, Prasad MNV (eds) Environmental adaptations and stress tolerance of plants in the era of climate change. Springer, New York, pp 29–56
Majumder J, Lodh R, Agarwala B (2013) Butterfly species richness and diversity in the Trishna Wild-life Sanctuary in South Asia. J Insect Sci 13(1):79
Malik K, Ahmad M, Ozturk M, Altay V, Zafar M, Sultana S (2021) Herbals of Asia – prevalent diseases and their treatments. Springer
Mandaville JP (1990) Flora of Eastern Saudi Arabia. Kegan Paul International, London, p 482
Maricle BR, Koteyeva NK, Voznesenskaya EV et al (2009) Diversity in leaf anatomy, and stomatal distribution and conductance, between salt marsh and freshwater species in the C4 genus *Spartina* (Poaceae). New Phytol 184:216–233
Masoodi KZ, Amin I, Mansoor S et al (2020) Chapter 11: botanicals from the Himalayas with anticancer potential – an emphasis on Kashmir Himalayas. In: Ozturk M, Egamberdieva D, Pešić M (eds) Biodiversity and biomedicine – our future. Academic Press, pp 189–234
Mercimek HV, Altay V, Ozturk M, Akçiçek E, Gucel S (2019) Chapter 15: an overutilised industrial crop tobacco – Nicotiana tabacum L. Case study from Turkey. In: Ozturk M et al (eds) Crop production technologies for sustainable use and conservation. Apple Academic Press, pp 365–394
Movsumova NV, Ibadullayeva SJ (2019) Halophytic plant diversity of Duzdag Area in Nakhchivan autonomous Republic: Azerbaijan. In: Hasanuzzaman M et al (eds) Ecophysiology, abiotic stress responses and utilization of halophytes. Springer, Singapore, pp 383–401
Mushtaq W, Mehdizade M, Siddiqui MB, Ozturk M, Jabran K, Altay V (2020) Phytotoxicity of above – ground weed residue against some crops and weeds. Pak J Bot 52(3):851–860
Nanduri KR, Hirich A, Salehi M, Saadat S, Jacobsen SE (2019) Quinoa: a new crop for harsh environments. In: Gul B et al (eds) Sabkha ecosystems, Tasks for vegetation science, vol VI. Springer, Cham, pp 301–333
O'Leary JW (1985) Halophytes Arizona Land and People 36(3):15
Obón C, Rivera D, Verde A, Alcaraz F (2021) Ethnopharmacology and medicinal uses of extreme halophytes. In: Grigore MN (ed) Handbook of halophytes: from molecules to ecosystems towards biosaline agriculture. Springer, Cham, pp 2707–2735
Omamt E, Hammes P, Robbertse P (2006) Differences in salinity tolerance for growth and water-use efficiency in some amaranth (*Amaranthus* spp.) genotypes. NZ J Crop Hortic Sci 34:11–22
Omar AS (2007) Vegetation of Kuwait: a comprehensive illustrated guide to the flora and ecology of the desert of Kuwait. Kuwait Institute of Scientific Research (KISR), Kuwait
Omar AS, Misak RF, Shahid S (2002) Sabkhat and halophytes of Kuwait. In: Barth HJ, Böer B (eds) Sabkha ecosystems, The Arabian Peninsula and adjacent countries, vol 1. Kluwer Academic, Dordrecht, pp 70–82
Osma E, Ozyıgıt II, Altay V, Serın M (2010) Urban vascular flora and ecological characteristics of Kadikoy district, Istanbul, Turkey. Maejo Int J Sci Technol 4(1):64–87
Ozturk M, Ozcelik H, Behcet L, Guvensen A, Ozdemir F (1995) Halophytic flora of Van Kale Basin Turkey. In: Khan MA, Ungar IA (eds) Biology of salt tolerant plants. Department of Botany, University of Karachi, Karachi, pp 306–315
Ozturk M, Guvensen A, Gork C, Gork G (2006) An overview of the coastal zone plant diversity and management strategies in the Mediterranean region of Turkey. In: Ozturk M, Waisel Y, Khan MA, Gork G (eds) Biosaline agriculture and salinity tolerance in plants. Birkhäuser, Basel, pp 89–100
Ozturk M, Guvensen A, Sakcali S, Gork G (2008) Halophyte plant diversity in the Irano-Turanian phytogeographical region of Turkey. In: Abdelly C, Ozturk M, Ashraf M, Grignon C (eds) Biosaline agriculture and high salinity tolerance. Birkhäuser, Basel, pp 141–155
Ozturk M, Altay V, Gucel S, Aksoy A (2012a) Aegean Grasslands as endangered ecosystems in Turkey. Pak J Bot 44:7–18

Ozturk M, Efe R, Çelık A et al (2012b) Comparative study on biogeography of protected and degraded habitats in Dilek Peninsula, Turkey. J Balkan Ecol 15(4):383–392
Ozturk M, Gucel S, Altay V, Altundağ E (2012c) Alliums, an underutilized genetic resource in the East Mediterranean. In: Proceedings of the 6th IS on Edible Alliaceae. Acta Hort. 969 (ISHS 2012), pp 303–309
Ozturk M, Guvensen A, Gucel S, Altay V (2013) An overview of the atmospheric pollen in Turkey and the Northern Cyprus. Pak J Bot 45:191–195
Ozturk M, Altay V, Gucel S, Guvensen A (2014a) Halophytes in the East Mediterranean-their medicinal and other economical values. In: Khan MA et al (eds) Sabkha ecosystems: Volume IV – cash crop halophyte and biodiversity conservation. Springer, Dordrecht, pp 247–272
Ozturk M, Dalgıç R, Guvensen A, Altay V, Gucel S (2014b) Honey-pollen-health: palinochemical analysis of honey from Turkey. In: Proceedings of the IS on medicinal plants and natural products. Acta Hort 1030, ISHS
Ozturk M, Altay V, Altundag E, Gucel S (2016) Halophytic plant diversity of unique habitats in Turkey: salt mine caves of Çankiri and Igdir. In: Khan MA, Ozturk M, Gul B, Ahmed MZ (eds) Halophytes for food security in dry lands. Elsevier, Oxford, pp 291–315
Ozturk M, Uysal I, Yucel E et al (2017a) Soil-plant interactions in the monumental plane trees (*Platanus orientalis*) grove – Canakkale-Turkey. J Environ Biol 38(6):1129–1137
Ozturk M, Altay V, Hakeem KR, Akçiçek E (2017b) Liquorice from botany to phytochemistry. SpringerBriefs in plant science. https://doi.org/10.1007/978-3-319-74240-3
Ozturk M, Altay V, Gonenç TM (2017c) Chapter 24: herbal from high mountains in the East Mediterranean. In: Bhojraj S et al (eds) Drug discovery from herbs-approaches and applications. NAM S & T Centre/DAYA Publishing House, New Delhi, pp 327–367
Ozturk M, Altay V, Gucel S, Altundağ E (2017d) Chapter 5: plant diversity of the drylands in Southeast Anatolia-Turkey: role in human health and food security. In: Ansari AA, Gill SS (eds) Plant biodiversity: monitoring, assessment and conservation. CABI, Wallingford, pp 83–124
Ozturk M, Altay V, Latiff A et al (2018a) Chapter 9: a comparative analysis of the medicinal pteridophytes in Turkey, Pakistan, and Malaysia. In: Ozturk M, Hakeem KR (eds) Plant and human health, vol 1. Springer, pp 349–390
Ozturk M, Gökler I, Altay V (2018b) Chapter 8: medicinal bryophytes distributed in Turkey. In: Ozturk M, Hakeem KR (eds) Plant and human health, vol 1. Springer, pp 323–348
Ozturk M, Altay V, Latiff A et al (2018c) Chapter 11: a comparative analysis of the medicinal plants used for diabetes mellitus in the traditional medicine in Turkey, Pakistan, and Malaysia. In: Ozturk M, Hakeem KR (eds) Plant and human health, vol 1. Springer, pp 409–461
Ozturk M, Altay V, Latiff A et al (2018d) Chapter 16: potential medicinal plants used in the hypertension in Turkey, Pakistan, and Malaysia. In: Ozturk M, Hakeem KR (eds) Plant and human health, vol 1. Springer, pp 595–618
Ozturk M, Altundağ E, Ibadullayeva SJ, Altay V, Aslanıpour B (2018e) A comparative analysis of medicinal and aromatic plants used in the traditional medicine of Iğdır (Turkey), Nakhchivan (Azerbaijan), and Tabriz (Iran). Pak J Bot 50(1):337–343
Ozturk M, Altay V, Altundağ E, Ibadullayeva SJ, Aslanipour B, Gönenç TM (2018f) Chapter 6: herbals in Iğdır (Turkey), Nakhchivan (Azerbaijan), and Tabriz (Iran). In: Ozturk M, Hakeem KR (eds) Plant and human health, vol 1. Springer, Cham, pp 197–266
Ozturk M, Altay V, Orçen N et al (2018g) Chapter 3: a little known and a little consumed natural resource: *Salicornia*. In: Ozturk M et al (eds) Global perspectives on underutilized crops. Springer, pp 83–108
Ozturk M, Altay V, Güvensen A (2019a) Sustainable use of halophytic taxa as food and fodder: an important genetic resource in southwest Asia. In: Hasanuzzaman M et al (eds) Ecophysiology, abiotic stress responses and utilization of halophytes. Springer, Singapore, pp 235–257
Ozturk M, Gucel S, Altay V et al (2019b) Clustering of halophytic species from Cyprus based on ionic contents, biochemical attributes and micro-nutrients. Phyton Int J Exp Bot 88(1):63–68

Ozturk M, Altay V, Kucuk M et al (2020) Preservation and ecology of a living relict shrub in South Caucasus as a eco-genetic heritage from Tertiary: *Epigaea gaultherioides* (Boiss. & Bal.) Takht. J Environ Biol 41:279–284

Ozturk M, Altay V, Efe R (2021a) Biodiversity, conservation and sustainability in Asia: volume I: prospects and challenges in West Asia and Caucasus. Springer. https://doi.org/10.1007/978-3-030-59928-7

Ozturk M, Altay V, Yarcı C, Yucel E, Kutbay HG, Kucuk M (2021b) Chapter 4: endangered swamp forests in Turkey – an ecological inventory, prospects and challenges. In: Ozturk M, Altay V, Efe R (eds) Biodiversity, conservation and sustainability in Asia: volume I: prospects and challenges in West Asia and Caucasus. Springer, pp 661–679. https://doi.org/10.1007/978-3-030-59928-7_4

Ozturk M, Altay V, Gönenç TM, Unal BT, Efe R, Akçiçek E, Bukhari A (2021c) An overview of olive cultivation in Turkey: botanical features, eco-physiology and phytochemical aspects. Agronomy 11:295

Ozturk M, Khan SM, Altay V, Efe R, Egamberdieva D, Khassanov F (2022) Biodiversity, conservation and sustainability in Asia: volume II: prospects and challenges in South and Middle Asia. Springer, Cham. https://doi.org/10.1007/978-3-030-73943-0

Ozyıgıt S, Altay V, Ozyığıt II, Yarcı C (2015) Vegetation ecology of the Princes' Islands, Istanbul-Turkey. J Environ Biol 36:113–120

Park HS, Sohn B (2010) Recent trends in changes of vegetation over East Asia coupled with temperature and rainfall variations. J Geophys Res Atmos 115:D14

Patel BB, Patel BB, Dave RS (2011) Studies on infiltration of saline–alkali soils of several parts of Mehsana and Patan districts of North Gujarat. J Appl Technol Environ Sanitation 1(1):87–92

Pleskanovskaya SA, Mamedova MA, Ashıralıyeva MA et al (2019) Chapter 2: Glycyrrhiza glabra (Liquorice) in Turkmenistan – medicinal and biological aspects. In: Ozturk M, Hakeem KR (eds) Plant and human health, – pharmacology and therapeutic uses, vol 3. Springer, Cham, pp 23–35

Quamar M, Ali SN, Pandita SK, Singh Y (2018) Modern pollen rain from Udhampur (Jammu and Kashmir), India: Insights into pollen production, dispersal, transport and preservation. Palynology 42(1):55–65

Raghunandan K, Basavarajappa S (2014) Floral hosts and pollen calendar of Asian giant honeybee, Apisdorsata Fabricius at Southern Karnataka, India. J Ecol Nat Environ 6(9):321–330

Rajpar MN, Ozturk M, Altay V et al (2020) Species composition of dry-temperate forest as an important habitat for wildlife fauna species. J Environ Biol 41:328–336

Rajpurohit KS (1980) Soil salinity and its role on phytogeography of western Rajasthan. PhD thesis, Jodhpur University, Jodhpur

Rancic D, Pecinar I, Acic S, Stevanovic ZD (2019) Morpho-anatomical traits of halophytic species. In: Hasanuzzaman M, Shabala S, Fujita M (eds) Halophytes and climate change: adaptive mechanisms and potential uses, CAB International, Wallingford, pp 152–178

Raza MM, Ullah S, Tariq A, Abbas T, Yousaf MM, Altay V, Ozturk M (2019) Alleviation of salinity stress in maize using silicon nutrition. Notulae Botanicae Horti Agrobotanici Cluj-Napoca 47(4):1340–1347

Reef R, Lovelock CE (2015) Regulation of water balance in mangroves. Ann Bot 115:385–395

Saenger P, Ragavan P, Sheue CR, López-Portillo J, Yong JWH, Mageswaran T (2019) Mangrove biogeography of the Indo-Pacific. In: Gul B et al (eds) Sabkha ecosystems, Tasks for vegetation science, vol VI. Springer, Cham, pp 379–400

Sandu G (1984). Solurile saline s, i alcalice din R.S.R. Ameliorarea lor. Ed. Ceres, Bucuresti

Sardar MF, Ahmad HR, Zıa-Ur-Rehman M et al (2018) Absorption of foliar-applied lead (Pb) in rice (*Oryza sativa* L.): a hydroponic experiment. Fresenius Environ Bull 27(10):6634–6639

Schreiber L, Hartmann K, Skrabs M, Zeier J (1999) Apoplastic barriers in roots: chemical composition of endodermal and hypodermal cell walls. J Exp Bot 50:1267–1280

Schulze ED, Beck E, Mçller-Hohenstein K (2005) Plant ecology. Springer, New York

Severoğlu Z, Altay V, Ozyıgıt II et al (2011) Some ecological characteristic and the flora of Gölcük District and its environs (Kocaeli-Turkey). Sci Res Essays 6(4):847–875
Sezer Y, Altay V, Ozyıgıt II, Yarcı C (2015) Woody vegetation of Şile and its environs (Istanbul/Turkey) and destruction of the area. J Environ Biol 36:163–170
Shabala S, Mackay A (2011) Ion transport in halophytes. Adv Bot Res 57:151–199
Shahrasbi S, Pirasteh-Anosheh H, Emam Y, Ozturk M, Altay V (2021) Elucidating some physiological mechanisms of salt tolerance in *Brassica napus* L. seedlings induced by seed priming with plant growth regulators. Pak J Bot 53(2):367–377
Shaltout KH, El-Halawagy EF, El-Garawany MM (1997) Coastal lowland vegetation of eastern Saudi Arabia. Biodivers Conserv 6:1027–1040
Shamsutdinov N (2008) Genetic resources and breeding of the halophytes in arid regions of the South Russia. In: Yokaş I et al (eds) Proceeding of biosaline agriculture & salinity. Muğla University Press, Muğla, pp 68–78
Shamsutdinov ZS, Shamsutdinov NZ (2002) Biogeocenotic principles and methods of degraded pastures phytomelioration in Central Asia and Russia. In: Ahmad R, Malik KA (eds) Prospects for saline agriculture. Tasks for vegetation science (TAVS, vol. 37), pp 29–35
Shamsutdinov NZ, Shamsutdinova EZ, Orlovsky NS, Shamsutdinov ZS (2017) Halophytes: ecological features, global resources, and outlook for multipurpose use. Her Russ Acad Sci 87(1):1–11
Shrivastava P, Kumar R (2015) Oil salinity: a serious environmental issue and plant growth promoting bacteria as one of the tools for its alleviation. Saudi J Biol Sci 22(2):123–131
Shuyu W, Zhe X (2004) The preliminary analysis of 5 coupled ocean-atmosphere global climate models simulation of regional climate in Asia. J Clim Environ Res 2004:2
Song J, Fan H, Zhao Y et al (2008) Effect of salinity on germination, seedling emergence, seedling growth and ion accumulation of a euhalophyte *Suaeda salsa* in an intertidal zone and on saline inland. Aquat Bot 88(4):331–337
Stenlid G (1956) Salt losses and redistribution of salts in higher plants. In: Encyclopedia of plant physiology, vol. 4, pp 615–637
Stevanovic DZ, Kresovic M, Pecínar I et al (2010) Distribution of the halophytic grass *Puccinellia limosa* (Schur.) Holomb. on salt affected soils in Serbia in relation to its main adaptive responses to salinity. Ekológia (Bratislava) 29(3):258–268
Stevanovic ZD, Acic S, Steševic D et al (2019) Halophytic vegetation in south-east Europe: classification, conservation and ecogeographical patterns. In: Halophytes and climate change: adaptive mechanisms and potential uses. CABI, Wallingford, pp 55–68
Tarakçı S, Altay V, Keskin M, Sümer S (2012) Beykoz ve çevresi (İstanbul)'nin kent florası. Karadeniz Fen Bilimleri Dergisi 2(7):47–66
Taşpınar K, Ozturk M, Altay V, Polat H (2019) Chapter 14: sugar beet – an overutilised ancient crop. In: Ozturk M et al (eds) Crop production technologies for sustainable use and conservation. Apple Academic Press, pp 321–363
Tug GN, Yaprak AE, Vural M (2011) An overview of halophyte plant diversity from salt Lake area Konya, Turkey. In: Ozturk M, Mermut AR, Çelik A (eds) Urbanisation, land use, land degradation and environment. Daya Publishing House, Delhi, pp 356–371
Tug GN, Yaprak AE, Vural M (2019) The floristical, ecological, and syntaxonomical characteristics of salt marshes and salt steppes in Turkey. In: Gul B et al (eds) Sabkha ecosystems, Tasks for vegetation science, vol VI. Springer, Cham, pp 413–446
ur Rehman T, Khan MA, Khan H, Shah WA, Altay V, Ozturk M (2020) Tillage, nutrients and weed control affects yield of sugar beet. Fresenius Environ Bull 29(5):3380–3387
von Sengbusch P (2003) Halophytes. Botanik Online University of Hamburg
Voronkova NM, Burkovskaya EV, Bezdeleva TA, Burundukova OL (2008) Morphological and biological features of plants related to their adaptation to coastal habitats. Russ J Ecol 39:1–7
Vuckovic R (1982) The novel association of the alliance Festucion pseudovinae Soo 1933. Acta Biologica Yugoslavia, Ekologija 17:15–23
Waisel Y (1972) Biology of Halophytes. Academic Press, New York

Walter H (1974) Die Vegetation der Erde in Oko-physiologischer Betrachtung. Vol. 2: Die gemasigten und arktischen Zonen (G. Fischer, Stuttgart, 1968; Progress) Moscow

Wang J, Ding J, Abulimiti A et al (2018) Quantitative estimation of soil salinity by means of different modeling methods and visible-near infrared (VIS_NIR) spectroscopy, Ebinur Lake Wetland, Northwest China. Peer-Rev J 6:1–24

Wanner A, Suchrow S, Kiehl K, Meyer W, Pohlmann N et al (2014) Scale matters: Impact of management regime on plant species richness and vegetation type diversity in wadden sea salt marshes. Agric Ecosyst Environ 182:69–79

Wucherer W, Breckle S-W, Dimeyeva L (2001) Flora of the dry sea floor of the Aral Sea. In: Breckle S-W, Veste M, Wucherer W (eds) Sustainable land use in deserts. Springer, Berlin, pp 38–51

Yaprak AE, Tug GN (2006) Halophytic endemics of Turkey. In: Ivanova D (ed) 2009. Plant, fungal and habitat diversity investigation and conservation. Proceedings of IV Balkan botanical congress, Sofia, 20–26 June 2006. Institute of Botany, Sofia, pp 234–238

Yarcı C, Altay V (2016) Kocaeli ve çevresindeki tarım alanlarının yabani ot florası. Erzincan Univ J Sci Technol 9(2):148–171

Yarcı C, Serin M, Altay V (2007) The segetal vegetation of Kocaeli Province (Turkey). Ekoloji 16(63):23–33

Yensen NP (2008) Halophyte uses for the twenty-first century. In: Khan MA, Weber DJ (eds) Ecophysiology of high salinity tolerant plants. Springer, Dordrecht, pp 367–396

Younessi-Hamzekhanlu M, Ozturk M, Altay V et al (2020) Ethnopharmacological study of medicinal plants from Khoy city of West Azerbaijan-Iran. Indian J Tradit Knowl 19(2):251–267

Youssef AM (2009) Salt tolerance mechanisms in some halophytes from Saudi Arabia and Egypt. Res J Agric Biol Sci 5:191–206

Yuan F, Leng B, Wang B (2016) Progress in studying salt secretion from the salt glands in recretohalophytes: how do plants secrete salt? Front Plant Sci 7:977

Zhao K, Song J, Feng G, Zhao M, Liu J (2011) Species, types, distribution, and economic potential of halophytes in China. Plant Soil 342(1):495–509

Chapter 2
Some Representative Medicinal Halophytes in Asia

2.1 *Aerva javanica* (Burm.f.) Juss. ex Schult

Family: Amaranthaceae
Synonym(s): *Achyranthes javanica* (Burm.f.) Pers.; *Iresine javanica* Burm.f.; *Illecebrum javanicum* (Burm.f.) L
English name: Kapok bush, Desert cotton
Flowering period: July–September
Distribution: Afghanistan, India, Iran, Iraq, Lebanon, Syria, Myanmar, Oman, Pakistan, Palestine, Saudi Arabia, and Yemen (Malik et al. 2021)
Altitude: Up to 1630 m
Habit: Shrub
Ecological type: Xerophyte
Habitat: On sandy or calcareous soils in semi-arid and arid regions
Phytochemicals: Steroids, triterpenes, lipids, flavonoids, tannins, saponins, alkaloids, sulphates, carbohydrates and glycosides (Mouhoub et al. 2018a)
Potential biological activity: The plant shows antioxidant, antiviral, antiplasmodial, antidiabetic, and anti-ulcer activities (Khan et al. 2012).
Toxicity: Subacute toxicity (Mouhoub et al. 2018a)

M. Ozturk et al., *Halophyte Plant Diversity and Public Health*,
https://doi.org/10.1007/978-3-031-21944-3_2

Botanical description: The plant is densely hairy, with thick foliage, woolly white to greyish in color, much-branched; erect shrub with the much-branched stem at the base; taproot is deep up to the same length as the aerial part of the plant, forming woolly tufts; leaves simple, alternate, shortly petiolate, oval, lance-shaped, hairy pinnately veined; two types of inflorescences, female (whitish with a purplish hue) and male (whitish with a golden yellow), both male and female inflorescence form woolly terminal panicles; each inflorescence is composed of 10 small racemes arranged on a 20 cm long twig; racemes and the main axis 6 cm long and laterally borne with very small flowers (2 mm) and pedicels of equal length; the female flower hypogynous, surrounded by three smooth transparent bracts, with 5 tepals and a nectariferous disc (10 nectariferous glands) in intrastaminal position; ovary with a short style and bifid feathery stigma; male flower consists of 5 tepals surrounded by bracts, stamens (5 fertile and 5 sterile or staminodes) attached to their basis, sterile ovary with a short style and small bifid stigma; fruit a pyxis, ovoid, blackish purple, surrounded by a persistent perianth (Fig. 2.1) (Palmer and Lally 2011; Mouhoub et al. 2018b).

Pollen features: The pollen grain is pantoporate (Müller and Borsch 2005).

Anatomical features: Mouhoub et al. (2018b) report that the plant organ surfaces are generally covered by glandular and non-glandular trichomes, developing from epidermal cells. The former are exclusively in uniseriate clusters, interrupted by secretory glands with unicellular star-shaped cells surrounded by a thick cellulose wall; the latter too are in uniseriate clusters and long. The epidermis is single layered in the leaves and stems, surface covered by numerous glandular and non-glandular branched trichomes, with typically star shaped tips of glandular ones. As the drought conditions increase, the number and length of trichomes also increase. A large number of submerged (pressed) stomata present on two faces of the epidermis (Fig. 2.2) (Mouhoub et al. 2018b).
Nazish et al. (2020) have reported that amphistomatic leaves observed in the plant, with anisocytic type of stomata. The trichome type has been reported stellate (Nazish et al. 2020).

Part used: Root, flower, stem, leaf, seed

Mode of utilization: Decoction, powder, extract, and paste

Route of administration: Oral, topical

Disease treated: Hematuria, kidney stones, dysuria, abortion, depression, bee sting, aphrodisiac, itching, cholera, phthisis, gonorrhea, Alzheimer's disease, and stomachic

Traditional uses: The plant is used in relieving the swelling and pain due to kidney stones. A decoction is used for dysentery, gonorrhea, and cutaneous infections (Khan et al. 2012).

Recipes:

- Flower extract is used orally for gonorrhea.
- Decoction of leaves is taken orally to cure stomachic.
- Dried stem powder is taken with milk to cure hematuria.
- The paste from leaves is used as poultice on bee sting to reduce pain.
- Seed powder is mixed with *Piper nigrum* seed powder and taken orally to cure cholera.

Potential economic significance: A good soil binder, drought-resistant, and helpful in controlling soil erosion. As per Global Sustainability Assessment System criteria, the plant has been categorized among the native species that could be used for urban landscaping particularly in desert climates (Global 2014; Phondani et al. 2015).

Fig. 2.1 *Aerva javanica* (Burm.f.) Juss. ex Schult

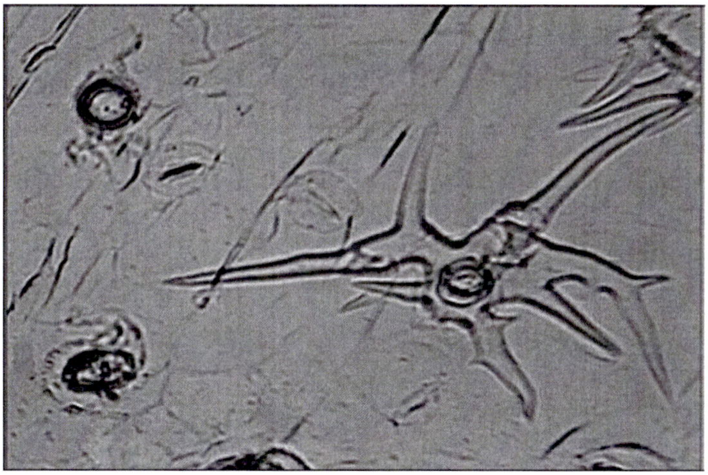

Fig. 2.2 Abaxial surface of *Aerva javanica* (Burm.f.) Juss. ex Schult

2.2 *Alhagi maurorum* Medik

Family: Fabaceae
Synonym(s): *Alhagi alhagi* (L.) Huth; *Hedysarum alhagi* L
English name: Camelthorn, Camelthorn-Bush, Caspian Manna, and Persian Mannaplant
Flowering period: July
Distribution: Afghanistan, Azerbaijan, China, India, Iran, Iraq, Jordan, Kazakhstan, Kyrgyzstan, Kuwait, Lebanon, Syria, Mongolia, North Caucasus, Pakistan, Palestine, Russia, Tajikistan, Transcaucasus, Turkiye, Turkmenistan, Uzbekistan, and West Himalayas (Malik et al. 2021)
Altitude: Up to 1000 m
Habit: Shrub
Ecological type: Hydrohalophyte
Habitat: The plant grows in different habitats such as inland salt marshes, dry salt marshes, disturbed lands, dry riverbeds and dunes, saline depressions, in low-lying steppe vegetation, on stabilized sand dunes in shady sites, on calcareous wastelands, in semi-arid regions, and sandy banks of some rivers on fresh alluvial deposits rich in calcium carbonate (Olama and Shehata 1993; Shaltout and Mady 1996; Abd el Ghani 2000).
Phytochemicals: Alkaloids, flavonoids, phenolics, anthraquinones, cardiac glycosides, sterols, coumarins, saponins, phlobatannins, tannins, terpenoids, vitamins, and fatty acids (Samejo et al. 2012; Ahmad et al. 2015)
Potential biological activity: The plant shows antibacterial, antipyretic, antidiuretic, antiulcer, antidiarrheal, anti-inflammatory, analgesic, antiproliferative, antioxidant, and antinociceptive activities (Samejo et al. 2012; Chakou et al. 2021).
Toxicity: Not reported
Botanical description: Rhizomatous, perennial shrub; leaves grayish-green and hairless, simple, entire oval to lance-shaped and arranged alternately; flowers pinkish purple to maroon, borne on short, spine-tipped branches; fruits reddish-brown, constricted between the seeds, with a short narrow beak at the end (Fig. 2.3) (Wazir et al. 2014).

Fig. 2.3 *Alhagi maurorum* Medik

Pollen features: The pollen grains of the species similar to the Ononis type; trizonocolporate, compressed oval in equatorial view, circular in polar view; exine <3 μm thick with a micro-reticulate surface; colpus membrane granulate; grain size <20 μm (Sekina and Moore 1995).

Anatomical features: The root cortex of the plant is large and collenchymatous surrounded by the epidermal layer. Numerous vascular bundles comprising of well-developed protoxylem and metaxylem are arranged radially. Metaxylem is lying towards the center while the protoxylem is towards the cortical zone. In the center, the parenchymatous pith retains moisture to help the xerophytic adaptation (Wazir et al. 2014).

The stem of the plant has a thick epidermis and a poorly developed cortex. Vascular bundles are arranged radially from the center towards the periphery. Xylem is well developed helping in the conduction of large amounts of water from the ground with no pith (Wazir et al. 2014).

Part used: Flower, leaf, gum

Mode of utilization: Extract, oil, powder, paste, decoction

Route of administration: Oral, topical

Diseases treated: Blood purifier, jaundice, respiratory diseases, asthma, aphrodisiac, laxative, demulcent, urinary diseases, syphilis, ringworms, measles, itching, abdominal diseases, malaria, blood purification, and sciatica

Traditional uses: Used in India as laxative, purgative, diuretic, for curing rheumatic pains / bilharzia and expectorant. The oil extracted from leaves is evaluated for curing rheumatism (Singh et al. 1990; El-Sayed et al. 1993). Seed decoction is used for curing kidney stones in Jordan, Israel, and Palestine (Fahmy 1963; Dafni and Lev 2002; www.cabi.org).

In folkloric local medicine, the plant has been used to treat nasal polyps, glandular tumors and bile duct related ailments, as diaphoretic, gastroprotective, diuretic, laxative, expectorant, antiseptic, anti-diarrheal and healing of wounds. The oil from leaves is used in the treatment of hemorrhoids and rheumatism. The flowers are used in the treatment of piles (James 2011; Shinwari et al. 2006; Ahmad et al. 2015).

Recipes:

- Flower extract is used to treat skin allergy.
- Leaf oil is used thrice a day to treat sciatica.
- Flower powder is taken before breakfast to cure glandular tumors.
- Extract of gum exuding from stem is taken orally for measles.
- Flower paste with *Anethum graveolens* leaf paste is taken twice a day against piles.
- Leaf decoction is used as ointment for itchy skin.

Potential economic significance: Widely cultivated as a highly palatable forage in China (Wang et al. 2001). In Afghanistan as well as India, it is used for making hay for small livestock and camels (Bhandari 1978; FAO 2002). In the desert regions, the plant is used locally as a source of fuel (Thalen 1979). In Indian arid regions, dry branches are widely used in making screens after wetting. These screens are evaluated for protection of the land against desiccating hot winds; sometimes planted for stabilization of sand dunes following the rains (www.cabi.org).

The stems and leaves provide a sugar exudate called "Manna" which is shaken from the bushes during flowering (Brandis 1972). Manna is sold as "torajabin" in India, being imported from Afghanistan and Iran, used for extracting mannitol, made into tablets and used in the cosmetic and pharmaceutical industries to produce laxatives, diuretics, and sweeteners (Maheshwari 1963; Goncharov et al. 2001).

2.3 *Artemisia scoparia* Waldst. & Kit

Family: Asteraceae
Synonym(s): *Artemisia capillaris* f. *elegans* (Roxb.) Pamp.; *Artemisia capillaris* var. *grandiflora* Pamp.; *Artemisia capillaris* f. *kohatica* (Klatt) Pamp.; *Artemisia capillaris* f. *myriocephala* Pamp.; *Artemisia capillaris* var. *scoparia* (Waldst. & Kit.) Pamp.; *Artemisia capillaris* f. *villosa* (Korsh.) Pamp.; *Artemisia capillaris* f. *williamsonii* Pamp.; *Artemisia elegans* Roxb.; *Artemisia gracilis* L'Hér. ex DC.; *Artemisia kohatica* Klatt; *Artemisia piperita* Pall. ex Ledeb.; *Artemisia scoparia* var. *heteromorpha* Kitag.; *Artemisia scoparia* f. *sericea* Kom.; *Artemisia scoparia* f. *villosa* Korsh.; *Artemisia scopariiformis* Popov; *Artemisia scoparioides* Grossh.; *Artemisia trichophylla* Wall. ex DC.; *Draconia scoparia* (Waldst. & Kit.) Soják; *Oligosporus scoparius* (Waldst. & Kit.) Less (Malik et al. 2021)
English name: Redstem wormwood
Flowering period: June–September
Distribution: Afghanistan, China, India, Iran, Iraq, Kazakhstan, Kyrgyzstan, Mongolia, Myanmar, North Caucasus, Pakistan, Russia, Saudi Arabia, Tajikistan, Tibet, Transcaucasus, Turkiye, Turkmenistan, Uzbekistan, West Himalayas, and West Siberia (Malik et al. 2021)
Altitude: Up to 2200 m
Habit: Herb
Ecological type: Psammohalophyte
Habitat: In the hot dry areas, on sandy clayey soil, wasteland, and rural areas, as well as stony steppes, and dry river beds (Boakye et al. 2017; Doležal et al. 2018; Malik et al. 2021).
Phytochemicals: Flavonoids, monoterpenes, sesquiterpenes, quercetin, quinic acid, 3-caffeoylquinic acid, 5-feruloylquinic acid and rutin, and essential oils (Malik et al. 2021)
Potential biological activity: Carminative, antipyretic, antiparasitic, anthelmintic, antiseptic, antispasmodic, antimicrobial, antibacterial, antiseptic, antipyretic, vasodilator, diuretic, anti-inflammatory, appetite, stimulating, digestive, fungicidal, emmenagogue, stomachic, vermifuge, vulnerary, and hypnotic, human periodontal infection (Malik et al. 2021).
Toxicity: Hepatotoxic and phytotoxic (Malik et al. 2021).
Botanical description: A biennial plant, branched above, 41–81 (−91) cm in length, hairy, glabrous, purple-brownish, and ting stem; lower stem has a long petiolate leaf 1.6 cm long, hairy, glabrous, ovate with orbicular lamina, 4.0–7.1 × 2.6–5.1 cm long; leaves on the upper stem sessile to sub-sessile, hairy, articulate, glabrous; heterogamous capitulate, many and short pediculate, 21–51 × 11–36 cm panicles with long primary and secondary branches; receptacles glabrous, conical; involucres 4–5-seriated, imbricate, glabrous, broad scarious hyaline; floret 11–13, yellowish; marginal florets 6–7, fertile, with 0.8 mm long, tube-like, petals double dentate; disc-florets 6–7, functional stamina, with 1.26 mm in length, and glandular corolla (Fig. 2.4) (Malik et al. 2021).

Fig. 2.4 *Artemisia scoparia* Waldst. & Kit

Pollen features: pollen shape: prolate, polar length: 19.2–24.5 µm, equatorial length: 11.2–15.7 µm (Ghahraman et al. 2007).
Anatomical features: Leaf symmetry: isobilateral 3 layered, stomata type: anomocytic, palisade cell wall: few depressions, midrib size: small (Noorbakhsh et al. 2008; Hussain 2020).
Part used: Leaves, bark
Mode of utilization: Decoction, extract
Route of administration: Oral
Disease treated: Skin infections and skin rashes
Traditional uses: Used for the treatment of burns, jaundice, hepatitis, and cure earache (Boakye et al. 2017; Malik et al. 2021).
Recipes: • 25 g of leaves are boiled in 400 mL water, filtrate used in the morning and evening for 14 days to treat skin infections (Malik et al. 2021). • 50 g of bark boiled in 200 mL of water for 30 minutes at low heat. 50 mL of hot extract given to children in the evening for 10 days for treating skin rashes (Malik et al. 2021).
Potential economic significance: The plant is generally known to be used as animal feed, but the shoots of plant are also used for making brooms, and fuel purposes (Ozturk et al. 2019; Hussain 2020). In addition, the uses such as ornamental and food has also been reported (Wiersema and León 1999).

2.4 *Atriplex halimus* L.

Family: Amaranthaceae
Synonym(s): *Atriplex halimoides* Tineo; *Atriplex halimus* var. *serrulata* (Pau) Alcaraz, Garre & Sánchez-Gómez; *Atriplex kataf* Ehrenb. ex Boiss.; *Atriplex serrulata* Pau; *Chenopodium halimus* (L.) Thunb.; *Obione halimus* (L.) G.L.Chu; *Schizotheca halimus* (L.) Fourr
English name: Mediterranean saltbush, sea orache, shrubby orache, and silvery orache
Flowering period: June–July
Distribution: Iraq, Iran, Jordan, Lebanon, Syria, Oman, Palestine, Pakistan, Saudi Arabia, and Turkiye
Altitude: Up to 1200 m
Habit: Shrub
Ecological type: Xerohalophyte
Habitat: In coastal sands, and saltmarshes
Phytochemicals: Alkaloids, phenolics, steroids, flavonoids, saponin, tannin and glycosides (Walker et al. 2014; Petropoulos et al. 2018).
Potential biological activity: Antibacterial, antidiabetic, antioxidant; anti-leishmanial; anti-multidrug resistance, immunomodulatory, and anticancer (Walker et al. 2014; Petropoulos et al. 2018).
Toxicity: Cytotoxicity (El-Aasr et al. 2016).
Botanical description: A sturdy erect branched shrub, 2–3 m, greyish white, peeling bark on older stem; leaves alternate, ovate-deltoid, or oblong ovate, up to 6 × 4 cm, silver grey, entire-repand lobed, subhastate, obtuse, mucronulate or emarginate, petiolate, petiole 0.3–1.0 cm; inflorescence terminal, paniculate, leafless; glomerules many-flowered forming congested lateral and terminal spikes; male flowers: perianth small, very few in number than female flowers; fruits many, condensed, often horizontally spreading; bracts thick-textured, fused at base, reniform, broad triangular to orbicular, 3.5–4 × 5–6 mm, entire or serrate dentate, without tubercles; seeds 1.5 mm, brown-dark brown (Fig. 2.5) (www.efloras.org).
Pollen features: Diameters: 22.33–30.03; exine thickness: 2.00–3.08 (Angelini et al. 2014).

Fig. 2.5 *Atriplex halimus* L.

Anatomical features: "Kranz" anatomy is a typical characteristic of the leaves, with a layer of bundle sheath cells surrounding each vascular bundle and radially arranged palisade cells, the bundle sheath is open (Shomer-Ilan et al. 1981; Walker et al. 2014); epidermis relatively thin and beneath it lies hypodermis – able to store water (Blumenthal-Goldschmidt and Poljakoff-Mayber 1968). The vesicular trichomes are present on the leaf surface, which is an important characteristic, particularly for stress tolerance (Smaoui et al. 2011; Walker et al. 2014). We find balloon-like hairs or bladder cells, 80–200 mm in diameter, with a surface coating of waxy material and attached to a stalk embedded in an epidermal cell (Smaoui et al. 2011; Walker et al. 2014).

Part used: Leaves

Mode of utilization: Boiling, decoction

Route of administration: Oral, topical

Disease treated: Heart diseases, diabetes, and rheumatism

Traditional uses: In diabetic patients, this plant is used in glycemic control and also for treating cardiological problems. It is also used to treat chest ailments, as a laxative, to cure stomach and muscular pains, treat intestinal worms and to regulate gall bladder excretions in the Arabian countries (Chikhi et al. 2014; Slama et al. 2020).

Recipes:

- A decoction of the plant leaves is drunk as a remedy for heart disease and diabetes (Said et al. 2002).
- The plant extract prepared with boiling water is added to bathwater to treat rheumatism (Said et al. 2002).

Potential economic significance: A recycling of plant mineral nutrients, incorporation of organic matter (root material and leaf litter) into the soil to improve soil physical features (such as water permeability) and provision of shelter and food for birds and mammals are the beneficial effects of planting this species (Henni and Mehdadi 2012; Walker et al. 2014).
The spore density of arbuscular mycorrhizal fungi increases in the desert soils in the presence of this species, also increasing the organic carbon and microbial biomass at soil depths down to 50 cm and augments the numbers of protozoa and nematodes (He et al. 2002; Pen-Mouratov et al. 2003; Rodriguez Zaragoza et al. 2005; Barness et al. 2009; Walker et al. 2014).
This plant is the best option for reclamation of degraded agricultural land in arid and semi-arid zones because it has a deep root system (Walker et al. 2014).
A possible application of the plant, suggested by its ability to stabilize soils physically and its tolerance of elevated concentrations of trace elements in the growth medium is in the remediation of trace element-contaminated soils (Walker et al. 2014).
The plant is also used as fodder to feed for camel, cattle, goat, and sheep (Walker et al. 2014; Ozturk et al. 2019).
The wood from this species has been used for centuries for heating and cooking, which is continuing even now in rural areas (Bouzid and Benabdeli 2011; Walker et al. 2014).

2.5 *Bassia indica* (Wight) A.J.Scott

Family: Amaranthaceae
Synonym(s): *Bassia joppensis* Bornm. & Dinsm.; *Kochia griffithii* Bunge ex Boiss.; *Kochia indica* Wight
English name: Indian bassia
Flowering period: July–October
Distribution: Afghanistan, Iran, Iraq, Kuwait, Pakistan, Palestine, Saudi Arabia, and West Himalayas (Malik et al. 2021).
Altitude: Up to 1500 m
Habit: Herb
Ecological type: Xerophyte
Habitat: Dry localities in the plains
Phytochemicals: Phytoecdysteroids, sterols, alkaloids, flavonoids, and saponins (Othman et al. 2021).
Potential biological activity: An extract of leaves is useful as an anti-inflammatory and analgesic (Qari et al. 2019). The plant has been reported to be used for improving cardiac activity, as well as anticancer, and antioxidant (Youssef 2013; Othman et al. 2021). In addition, the plant shows anti-fungal activity (Javed et al. 2018).
Toxicity: Resinous alkaloid, isolated from alcoholic extract of the plant, showed nicotinic action on autonomic ganglion and neuromuscular junction of voluntary muscles (Khare 2007).
Botanical description: Annual or biennial herb, up to 2 m, softly villous or pubescent, stem pale, brown to yellowish, with many spreading branches from the base and distant leaves; leaves 5–15(−30) × 1–5 mm, lanceolate, oblong or linear, entire, acute or sub-acuminate at apex, cuneate-sessile or subsessile, soft hairy to sub-glabrous; flowering branches usually whitish 1–3-flowered clusters arranged in loose shape, scattered, leafy spikes; bracts linear, leaf-like, longer than clusters, softly hairy; perianth ± woolly, connivent, up to 3 mm in fruit; shortly winged or not, with the variable wing shape at the back, rudimentary (or suppressed) to triangular-ovate to sub-orbicular, up to 1 mm long in diameter, scarious; seed c. 2 mm in diameter (Fig. 2.6) (www.efloras.org).
Pollen features: Pollen grain diameter: 21 μm, width of aperture: 1.25 μm, exine: smooth, small pollen (Turki et al. 2006).
Anatomical features: Nazish et al. (2020) have reported that amphistomatic leaves found on the plant, with anomocytic type of stomata. In addition, various types of trichomes have been reported, including glandular, non-glandular, and conical (Fig. 2.7) (Nazish et al. 2020). In the young stem cross-section is terete having slightly ridged margins 2.0–2.3 mm in diameter. It is hairy with multicellular hairs which are uniserate, papillate and smooth, with swollen basal cell and acute apical cell. The epidermis is of one layer, covered with wavy cutine, with more or less pentagonal shaped cells. Cortex is wide, has 2–3 outer layers of pentagonal-hexagonal cells collenchyma, 3–4 layers of irregularly flattened parenchyma follow it with well-defined sand crystals; endodermis well defined, pericycle as sclerenchymatous patches facing vascular bundles; vascular cylinder composed of 16–18 vascular bundles which are equidistant from the center of the stem, separated by medullary rays; phloem 2–4 layers, cambium 1–3 layers; xylem vessels angular, arranged in rows; pith 1200 μm in width, having pentagonal - polygonal thick-walled parenchyma, with sand crystals (Turki et al. 2006).
Part used: Leaf, flower, seed
Mode of utilization: Decoction, paste, raw
Route of administration: Topical
Disease treated: Heart tonic, toothache, contraceptive, bug bites, urinary problems, swollen feet, kidney disorders, urine inflammation, and blood formation.
Traditional uses: Traditionally, the plant is used to treat renal and rheumatic diseases (Qari et al. 2019).

Recipes:

- Leaf decoction is applied externally to treat sunstroke.
- Flowers are chewed to stop tooth bleeding.
- Leaf paste is applied as poultice to reduce pain on bug bites.
- Seed decoction is applied on swollen feet to reduce swellings.

Potential economic significance: The plant is considered to have potential as green fodder on a global scale (Inam-ur et al. 2011; Hashem et al. 2019; Qari et al. 2019). It grows on saline soil and provides a dense cover for the soil surface, aiding in soil conservation and management (Shaltout and El-Beheiry 2000; Hand 2003); these traits have also been used for the detection of microbial habitats in certain locations (Hashem et al. 2015). It is adapted to abiotic stress and has been used in the repair of desert ecosystems, phytoremediation of saline soils and for livestock grazing on lands affected by salinity (Hashem et al. 2019; Qari et al. 2019).
People have used the plant for nutrition, house construction, manufacture of clothes, medication, cosmetics, ceremonial use, and also in magic (Qari et al. 2019).
In addition, the plant has a high content of sugars, lipids, and proteins, making it suitable as a safe fodder for animals, such as cattle, and as a natural organic fertilizer (Nafea 2017; Qari et al. 2019).

Fig. 2.6 *Bassia indica* (Wight) A.J.Scott

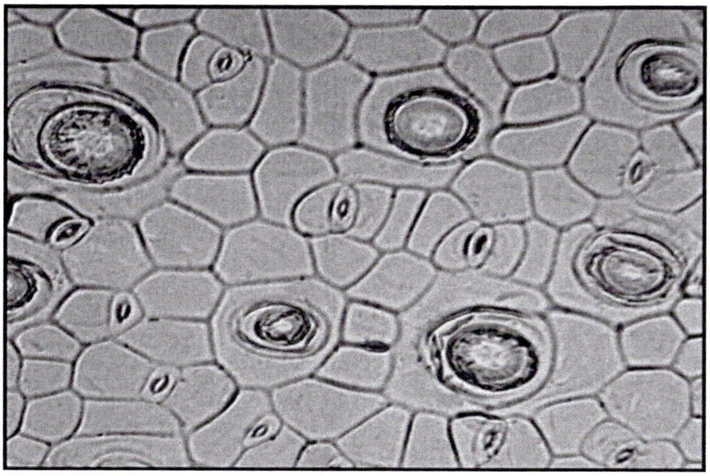

Fig. 2.7 Abaxial surface of *Bassia indica* (Wight) A.J.Scott

2.6 *Cakile maritima* Scop.

Family: Brassicaceae
Synonym(s): *Bunias cakile* L.; *Cakile cakile* (L.) H.Karst.; *Crucifera cakile* (L.) E.H.L.Krause; *Rapistrum cakile* (L.) Crantz; *Rapistrum maritimum* (Scop.) Bergeret
English name: Sea rocket
Flowering period: June–August
Distribution: Iran, Lebanon, North Caucasus, Palestine, Russia, Syria, Transcaucasus, Turkiye
Altitude: Sea level – 40 m
Habit: Herb
Ecological type: Psammohalophyte
Habitat: Sandy and shingly shores.
Phytochemicals: Alkaloids, phenolics, ascorbic acid, flavonoids, phytosterols, terpenoids, triterpenes, coumarins, sulpher glycosides and fatty acids (Ksouri et al. 2012; Petropoulos et al. 2018; Placines et al. 2020).
Potential biological activity: Antioxidant, antibacterial, antiscorbutic, anti-inflammatory, and anti-proliferative activities (Petropoulos et al. 2018; Fuochi et al. 2019).
Toxicity: Cytotoxicity (Tawfik et al. 2021).
Botanical description: Annual herb with a long taproot and a ± prostrate or ascending stem up to c. 40 cm; lower leaves 2–11 cm, pinnate or pinnatifid, the lobes oblong or more rarely toothed; upper leaves sessile and entire; petals 11x4 mm, white, lavender or rose; fruiting pedicels thick, 2–5 mm, erect-spreading; fruit 20–25 mm, glabrous, of 2 unequal, 1-seeded members, rarely the basal member 2-seeded; each member indehiscent; the upper oblong-ovate, c. 15 mm, the lower turbinate, c. 5 mm, often with 2 lateral horns, hastate at the joint (Fig. 2.8) (Davis 1965).

Fig. 2.8 *Cakile maritima* Scop.

Pollen features: Pollen grain: trizonocolpate; pollen sculpturing: macro-reticulate; pollen grain size: 54 µm (Sekina and Moore 1995).

Anatomical features: Jianu et al. (2014) have reported that the root has a secondary structure, due to the phellogen and cambium activity; stem has one-layered epidermis, covered by thick cuticle, a differentiated cortex and large number of collateral vascular bundles in upper part. The leaves are amphistomatic, leaf lobes have a homogenous mesophyll. The sclerenchymatous fibers in the root and collenchyma tissue in the stem represent the mechanical tissue (Jianu et al. 2014).

Part used: Leaves

Mode of utilization: Eaten fresh

Route of administration: Oral

Disease treated: Antiscorbutic, diuretic, and purgative

Traditional uses: The plant has been used for treating some chronic diseases such as cancer, inflammation, atherosclerosis, and Alzheimer (Liguori et al. 2018; Tawfik et al. 2021). The plant is also used as antidiuretic, antiscorbutic, headaches, and as purgative (Arbelet-Bonnin et al. 2019).

Recipes: The leaves are consumed fresh, used as antiscorbutic, diuretic, and purgative (Davy et al. 2006).

Potential economic significance: Arbelet-Bonnin et al. (2019) have reported that this plant is a good halophytic genetic model, possessing a small diploid genome, short life cycle and produces large number of seeds. It is a promising species because of its large geographical and ecological amplitude, economic potential, as well as ability to produce numerous secondary compounds. It is valuable as an oilseed and energy crop and cultivation on saline marginal lands is of practical significance in the context of the necessary development of future biosaline agriculture (Arbelet-Bonnin et al. 2019).

2.7 *Capparis decidua* (Forssk.) Edgew.

Family: Capparaceae
Synonym(s): *Capparis aphylla* Roth; *Capparis sodada* R.Br.; *Maerua linearis* Pax; *Niebuhria linearis* DC.; *Sodada decidua* Forssk.
English name: Caper berry
Flowering period: March–April
Distribution: India, Iran, Oman, Pakistan, Palestine, Saudi Arabia, West Himalayas, Yemen
Altitude: 300–1200 m
Habit: Shrub
Ecological type: Xerophyte
Habitat: In dry, hot, sandy desert areas, and arid regions.
Phytochemicals: Alkaloids, phenolics, flavonoids, flavonols, terpenoids, sterols, polyamines, and fatty acids (Nazar et al. 2020).
Potential biological activity: The plant has been reported to show antidiabetic, anthelmintic, antibacterial, antifungal, analgesic, anti-nociceptive, antirheumatic, hypolipidemic, antiatherosclerotic, anti-tumor, antigiardial, antioxidant, anti-inflammatory, hepatoprotective, and anticonvulsant activities (Nazar et al. 2020).
Toxicity: Cytotoxic (Nazar et al. 2020).
Botanical description: A shrub of dense tufts, 4–5 m high; possesses many green vine-like apparently leafless branches, hanging in bundles; branches and twigs glossy dark green and bark turns whitish-grey with age; small, light brown spines occur in pairs on the twigs at each node; leaves very minute (2 mm long) and emerge on young shoots; very short living and the plant looks leafless most of the time; flowers pink, red-veined, emerge in small groups along with the leafless shoots at the axils of the spines; fruit small many-seeded ovoid or sub-globulous, slightly mucronate and becomes blackish when dry; the seeds 2–5 mm diameter, and embedded in the pulp; large size fruits many (>30), small seeds while smaller fruits together with 1–2 large sized seeds (Fig. 2.9) (Satyanarayana et al. 2008; Mahla et al. 2013; Malik et al. 2021).
Pollen features: Pollen grains prolate, tricolporate, trilobed, size: polar axis P 20.16 ± 0.28 μm, and equatorial diameter E 13.9 ± 0.45 μm; exine 1.47 ± 0.07 μm thick (Perveen and Qaiser 2001).

Fig. 2.9 *Capparis decidua* (Forssk.) Edgew.

Anatomical features: According to Rathee et al. (2009), the transverse section of the stem reveals the presence of cuticle, epidermis, starch sheath, pericycle, phloem, cambium, xylem, medullary ray, and pith.

Part used: Fruit, stem, root, flower, and bark

Mode of utilization: Raw, paste, powder, decoction

Route of administration: Oral, topical

Disease treated: Digestive disorders, blood purifier, aphrodisiac, impotence, tuberculosis, teeth disorders, arthritis, Alzheimer's disease, skin disorders, piles, painful menstruation, high blood sugar, bone fractures, intestinal worms, sexual stimulant, increased perspiration, and joints swelling.

Traditional uses: The plant is used for the treatment of rheumatism, asthma, cough, lumbago, toothache, pyorrhea, dysentery, liver infections, diarrhea, febrifuge, cardiac troubles, constipation, ulcer, piles, renal disorders, and skin diseases (Nazar et al. 2020).

Recipes:

- Fruit is eaten raw for digestive disorders and blood purification.
- The twig paste is applied on teeth for shining.
- Flower powder is used dermally to treat pimples.
- Fruit is used to make local dish pickle which prevents from gastric pain.
- Stem is used as toothpick to cure toothache.
- Paste of slender shoots is applied externally on skin blisters as cure.
- Decoction of root bark is applied externally to cure swollen joints.
- Flowers are eaten in raw form with milk against intestinal worms.

Potential economic significance: A rich source of basic mineral nutrients with important nutraceutical values. The calcium and potassium content is considerably high, and is thus valued both for livestock feed and human food. The flowers and fruits are a potential source of important electrolytic minerals, in particular iron and zinc. Both are present in the plant supporting its potential use in the mineral deficiency in human diet (Gull et al. 2015; Nazar et al. 2020).

The plant is extremely resistant to drought, salinity, and soil erosion, and tolerates frost, resulting in a plant with excellent adaptations to arid conditions (Ozcan 2005; Ghangro et al. 2015; Nazar et al. 2020).

Jain (1994) and Mahla et al. (2013) have reported that the plant is used as firewood and for preparing charcoal because its wood has good fuelwood features such as calorific value, density, silica, carbon, ash-biomass ratio, moisture, volatile matter and fuelwood index. The wood is also used to make water pipes and water troughs as it is very hard. The timber is also hard, heavy and termite resistant, therefore the strength and durability of its wood are suitable for making small beams, rafts, knees of boats, tool handles, cartwheels, axles, and even combs; also used for making huts and fences.

In the semi-desert and desert areas, it is used in landscape gardening, afforestation and reforestation as it prevents soil erosion (Bangarwa 2008; Mahla et al. 2013); reduces wind erosion and alkalinity, improves soil fertility by increasing organic carbon and available N, P and K; regarded as the best species for shelter belts to control sand movement. It has been reported to play an important role in sand dune stabilization in the Indian Thar Desert (Pandey and Rokad 1992; Mahla et al. 2013; Malik et al. 2021).

2.8 *Cenchrus ciliaris* L.

Family: Poaceae

Synonym(s): *Cenchrus anjania* Buch.-Ham. ex Wall.; *Cenchrus bulbosus* Fresen. ex Steud.; *Cenchrus ciliaris* subsp. *ibrahimii* Chrtek & Osb.-Kos.; *Cenchrus glaucus* C.R.Mudaliar & Sundararaj; *Cenchrus longifolius* Hochst. ex Steud.; *Cenchrus melanostachyus* A.Camus; *Cenchrus mutabilis* Wight ex Hook.f.; *Cenchrus rufescens* Desf.; *Pennisetum cenchroides* Rich.; *Pennisetum ciliare* (L.) Link; *Pennisetum distylum* Guss.; *Pennisetum incomptum* Nees ex Steud.; *Pennisetum longifolium* Fenzl ex Steud.; *Pennisetum oxyphyllum* Peter; *Pennisetum panormitanum* Lojac.; *Pennisetum petraeum* Steud.; *Pennisetum polycladum* Chiov.; *Pennisetum rangei* Mez; *Pennisetum rufescens* (Desf.) Spreng.; *Pennisetum teneriffae* Steud.

English name: Buffel-grass; Foxtail grass

Flowering period: July–August

Distribution: Afghanistan, Bangladesh, India, Iran, Iraq, Kuwait, Lebanon, Oman, Pakistan, Palestine, Saudi Arabia, Syria, and Yemen (Malik et al. 2021).

Altitude: Up to 1000 m

Habit: Herb

Ecological type: Psammohalophyte

Habitat: Arid and semi-arid habitats

Phytochemicals: Alkaloids, anthraquinones, flavonoids, glycosides, steroids, phenols, saponins and tannins (Aleem and Janbaz 2017).

Potential biological activity: The plant showed the presence of antifungal, antibacterial, anthelmintic, antischistosomal, antiemetic, spasmolytic, antidiarrheal, dysmenorrheal, and uterine relaxing activities (Aleem and Janbaz 2017).

Toxicity: Cytotoxic (Alothman et al. 2018), and phytotoxic (Hussain et al. 2010).

Botanical description: Perennial, often forming mats or tussocks; culms 10–150 cm high, ascending, wiry; leaf blades 3–25 cm long, 2–10 mm wide; panicle cylindrical to ovoid, 2–12 cm long, 10–26 mm wide, grey, purple or straw-colored, the rachis angular and puberulous; involucre elongate, 6–16 mm long; inner bristles greatly exceeding the spikelets, one of them longer and stouter than the rest, at least the longest somewhat flattened at the base, connate only at the base to form a disc 0.5–1.5 mm in diameter, sparsely or densely ciliate below, grooved on the face or not, filiform above, flexuous, often wavy, antrorsely scaberulous; outer bristles filiform; spikelets 2–4 per burr, 2–5.5 mm long, acutely lanceolate; glumes distinct, acute, the lower as long as the spikelet (Fig. 2.10) (www.efloras.org).

Pollen features: Polar diameter: 45–55 μm; equatorial diameter: 42.50–47.50 μm; exine thickness: 0.4–1.1 μm; exine sculpturing: verrucate-reticulate; shape: spheroidal; aperture condition: monoporate; position of pore: endoporus; pore orientation: sunken (Shaheen et al. 2011; Ullah et al. 2021).

Fig. 2.10 *Cenchrus ciliaris* L.

Anatomical features: The salt stress has been reported to induce anatomical modifications in the root, stem and leaf (Wasim and Naz 2020). These help in the survival of this plant under adverse environmental conditions. The salinity increases the thickness of endodermis and sclerenchyma tissues in the roots thereby preventing the water loss from the root surface and increases the number of parenchyma cells in the pith and cortex region, improving the root water storage capacity. Salinity contributes to the epidermis and sclerenchyma thickness in the stems along with an increase in the number of vascular bundles and its area. These features improve the conduction of water and solute, increase succulence ability and prevent water loss. In the leaf and leaf sheath stomatal density decreases but the area is increased by bulliform cells. A rich density of vesicular hairs and trichomes looks like an essential feature for water conservation and salt excretion (Fig. 2.11) (Wasim and Naz 2020).

Part used: Whole plant

Mode of utilization: Powder, infusion, extract, paste

Route of administration: Oral, topical

Disease treated: Increase lactation, body pain, emollient, kidney stones, wounds, tumors, sores, blood diseases, diuretic, skin allergy, hepatic disorders, psoriasis, polyuria, fever, and stomach acidity.

Traditional uses: The plant is used as a galactagogue, emollient, purgative and diuretic, also in renal colic, tumor, wound, sores, menstrual disorders, and urinary tract infections (Mahmood et al. 2011; Ashraf et al. 2013; Aleem and Janbaz 2017).

Recipes:

- Dried seed powder is used as diaphoretic.
- Dried leaf powder is taken with water for fever.
- Infusion of whole plant is used orally for kidney pain.
- Seed extract is taken orally to treat hepatic ailments.
- Seed powder is taken after meal to reduce stomach acidity.
- Paste of stem is applied dermally to cure psoriasis.
- Leaf infusion with *Piper nigrum* seed powder is taken orally to cure polyuria.

Potential economic significance: In India, sometimes the plant seeds are added to the millet for making bread (Quattrocchi 2006). This plant is also economically important as fodder for domestic animals. Its tolerance to drought and overgrazing lead to an increased production compared to native grasslands, especially in marginal environments (Martin et al. 1995; Marshall et al. 2011; www.cabi.org).
In Australia, it has been used to restore productivity on degraded lands around the mining sites (Bisrat et al. 2004; www.cabi.org).

Fig. 2.11 Abaxial leaf surface of *Cenchrus ciliaris*

2.9 *Chenopodium album* L.

Family: Amaranthaceae

Synonym(s): *Anserina candidans* (Lam.) Montandon; *Atriplex alba* (L.) Crantz; *Atriplex viridis* (L.) Crantz; *Blitum viride* (L.) Moench; *Botrys alba* (L.) Nieuwl.; *Botrys alba* var. *pauper* Lunell; *Botrys pagana* (Rchb.) Lunell; *Chenopodium agreste* E.H.L.Krause; *Chenopodium album* f. *cymigerum* (W.D.J.Koch) Aellen; *Chenopodium album* f. *dubium* Arlt & Jüttersonke; *Chenopodium album* f. *glomerulosum* (Rchb.) Aellen; *Chenopodium album* f. *heterophyllum* Wang-Wei & P.Y.Fu; *Chenopodium album* f. *lanceolatum* (Muhl. ex Willd.) Schinz & Thell.; *Chenopodium album* f. *leiospermum* Kuntze; *Chenopodium album* f. *opuliforme* Aellen; *Chenopodium album* f. *ovalifolium* Aellen; *Chenopodium album* f. *paucidentatum* Aellen; *Chenopodium album* f. *pseudozschackei* Aellen; *Chenopodium album* f. *spicatum* (W.D.J.Koch) Aellen; *Chenopodium album* subsp. *bernburgense* Murr; *Chenopodium album* subsp. *collinsii* Murr; *Chenopodium album* subsp. *densifoliatum* Ludw. & Aellen; *Chenopodium album* subsp. *hastatum* (C. Klinggr.) Graebn.; *Chenopodium album* subsp. *pedunculare* (Bertol.) Murr; *Chenopodium album* subsp. *pseudopulifolium* (Scholz) Murray; *Chenopodium album* var. *album; Chenopodium album* var. *candicans* Moq.; *Chenopodium album* var. *coronatum* Beauge; *Chenopodium album* var. *cymigerum* W.D.J.Koch; *Chenopodium album* var. *dacoticum* Aellen; *Chenopodium album* var. *desertorum* Kuntze; *Chenopodium album* var. *glomerulosum* (Rchb.) Peterm.; *Chenopodium album* var. *lanceolatum* (Muhl. ex Willd.) Coss. & Germ.; *Chenopodium album* var. *laxiflorum* Wang-Wei & P.Y.Fu; *Chenopodium album* var. *missouriense* (Aellen) Bassett & Crompton; *Chenopodium album* var. *paganum* (Rchb.) Syme; *Chenopodium album* var. *polymorphum* Aellen; *Chenopodium album* var. *pseudopulifolium* (Scholz) Abrom.; *Chenopodium album* var. *spicatum* W.D.J.Koch; *Chenopodium album* var. *stevensii* Aellen; *Chenopodium album* var. *subaphyllum* (Phil.) Reiche; *Chenopodium album* var. *viride* (L.) Moq.; *Chenopodium album* var. *viride* (L.) Pursh; *Chenopodium bernburgense* (Murr) Druce; *Chenopodium bicolor* Bojer ex Moq.; *Chenopodium borbasiforme* (Murr) Druce; *Chenopodium borbasii* F.Murr; *Chenopodium browneanum* Schult.; *Chenopodium candicans* Lam.; *Chenopodium catenulatum* Schleich. ex Steud.; *Chenopodium concatenatum* Willd.; *Chenopodium diversifolium* var. *montuosum* F.Dvořák; *Chenopodium elatum* Shuttlew. ex Moq.; *Chenopodium glomerulosum* Rchb.; *Chenopodium laciniatum* Roxb.; *Chenopodium lanceolatum* Muhl. ex Willd.; *Chenopodium lanceolatum* R.Br.; *Chenopodium lanceolatum* f. *opizii* F.Dvořák; *Chenopodium lanceolatum* f. *sessiliflorum* F.Dvořák; *Chenopodium lanceolatum* var. *antiquitum* F.Dvořák; *Chenopodium leiospermum* DC.; *Chenopodium lobatum* (Prodán) F.Dvořák; *Chenopodium missouriense* Aellen; *Chenopodium missouriense* var. *bushianum* Aellen; *Chenopodium neglectum* Dumort.; *Chenopodium neoalbum* F.Dvořák; *Chenopodium opulaceum* Neck.; *Chenopodium ovalifolium* (Aellen) F.Dvořák; *Chenopodium paganum* Rchb.; *Chenopodium paucidentatum* (Aellen) F.Dvořák; *Chenopodium pedunculare* Bertol.; *Chenopodium probstii* Aellen; *Chenopodium probstii* f. *lanceolatum* Aellen; *Chenopodium probstii* f. *parvoangustifolium* Aellen; *Chenopodium pseudoborbasii* f. *aellenii* F.Dvořák; *Chenopodium pseudoborbasii* f. *albiforme* F.Dvořák; *Chenopodium pseudoborbasii* f. *borbasiiforme* F.Dvořák; *Chenopodium pseudoborbasii* f. *longipedicellatum* F.Dvořák; *Chenopodium pseudoborbasii* f. *ramosum* F.Dvořák; *Chenopodium riparium* Boenn. ex Moq.; *Chenopodium serotinum* Ledeb.; *Chenopodium subaphyllum* Phil.; *Chenopodium superalbum* F.Dvořák; *Chenopodium superalbum* f. *kuehnii* F.Dvořák; *Chenopodium viride* L.; *Chenopodium viridescens* (St.-Amans) Dalla Torre & Sarnth.; *Chenopodium vulgare* Gueldenst. ex Ledeb.; *Chenopodium vulpinum* Buch.-Ham.; *Chenopodium zobelii* Murr ex Asch. & Graebn.; *Chenopodium zobelii* f. *hircinifolium* Aellen; *Chenopodium zobelii* f. *multidentatum* Aellen; *Chenopodium zobelli* A. Ludw. & Aellen; *Chenopodium* × *borbasioides* f. *hircinifolium* (Aellen) Hyl.; *Chenopodium* × *densifoliatum* (Ludw. & Aellen) F.Dvořák (www.powo.science.kew.org).

English name: Lamb's quarters, Melde, White goosefoot

Flowering period: July–October

Distribution: Cosmopolitan

Altitude: Up to 3600 m	
Habit: Herb	
Ecological type: Xerohalophyte	
Habitat: The plant is distributed alongside the roads, riverbanks, pastures, strips of uncultivated land and on wasteland.	
Phytochemicals: Phenols, lignins, alkaloids, flavonids, glycosides, saponins, essential oils, trigonelline, chenopodine, potassium, and vitamin C (Singh et al. 2010b; Choudhary and Sharma 2014).	
Potential biological activity: According to Poonia and Upadhayay (2015), pharmacologically this plant is a good candidate for the development of treatments for muscular spasms and pain. Sikarwar et al. (2017) have reported that the leaf extracts of this plant (methanolic and aqueous) have shown antilithiatic effects on experimentally induced urolithiasis in rats compared to a standard antilithiatic agent, cystone. This plant also shows significant anthelmintic activity against cyathostomins, an important gastrointestinal nematode infecting equids. The plant has considerable potential as an anthelmintic forage or feed supplement (Peachey et al. 2015; www.cabi.org).	
Toxicity: Due to the amounts of oxalic acid and nitrate that the plant contains, a high consumption is not recommended. The consumption of 400 g of fresh leaves of the plant could surpass the minimal lethal dose of oxalic acid for an adult (Guil et al. 1996, 1997).	
Botanical description: Annual, 10–150 cm, usually erect, variously branched, ± grey farinose; stems yellowish to green, green-striated, sometimes reddish or with red spots at leaf axils; lower and medium leaves petiolate, blade usually 2–6(−10) cm, variously trullate, rhombic-ovate to lanceolate, clearly longer than broad, base narrowly to broadly cuneate, margins irregularly serrate to entire, often somewhat 3-lobed, teeth mostly acute, often unequal in size; uppermost leaves lanceolate, usually entire; inflorescence a variable spiciform or cymosely branched panicle, mostly terminal; perianth segments 5, dorsally keeled; perianth falling with fruit; pericarp thin, ± adherent; seeds horizontal, black, 1.1–1.5 mm in diameter, somewhat ovate, margin weakly acute; testa with faint radial striae, otherwise almost smooth (Fig. 2.12) (www.efloras.org).	
Pollen features: Pollen type: apolar; pollen shape: spheroidal; pollen ornamentation: micro-echinate; polar diameter: 12.75–18.25 μm; exine thickness: 1.0–1.5 μm (Nazish et al. 2019).	

Fig. 2.12 *Chenopodium album* L.

Anatomical features: According to Nazish et al. (2020), amphistomatic leaves have been observed in the plant, with paracytic type of stomata. In addition, various types of trichomes have been reported, including glandular, non-glandular, and conical (Nazish et al. 2020).

Part used: Seed, leaf, root

Mode of utilization: Raw, cooked, oil, extract, infusion, juice, and powder

Route of administration: Oral, topical

Disease treated: Constipation, jaundice, aphrodisiac, abdominal pain, eye disorders, anthelmintic, laxative, sciatica, hepatitis B, gastric disorders, improve appetite, blood dysentery, sexual tonic, skin warts, and sleep apnea.

Traditional uses: Hindu physicians have been recommending this plant for the treatment of hepatic disorders and splenic enlargement (Chopra et al. 1958). Zulu tribals have been using the finely powdered leaf of this plant as a dusting powder to allay irritation of the external genitalia of children (Watt and Breyer-Brandwijk 1962). For appetite improvement he seeds are being used traditionally. It is also used as anthelmintic, laxative, aphrodisiac, tonic as well as to treat biliousness, stomach pains, eye and throat problems, piles, and diseases of blood, heart and spleen (Jansen 2004).

Recipes:

- Seeds are eaten in raw with water to get relieve from constipation.
- Seed oil is used orally to kill intestinal worms.
- Root extract is used orally to treat hepatitis B.
- Leaves are eaten as salad for gastric problems.
- Leaf extract consumed in the morning improves appetite.
- Leaf infusion is taken orally to cure sciatica.
- Root juice is used orally to cure blood dysentery.
- Seed powder is taken orally with water to cure sleep apnea.

Potential economic significance: The leaves and tender branches of the plant can be used as a vegetable at global level; also used in the production of curd, locally known as Raita in India (Maheshwari 1963). Young shoots are boiled and eaten often with other vegetables, also dried and stored for later use (Jansen 2004; www.cabi.org).

Previously seeds of the plant were harvested all over Europe, dried and ground into flour for making bread, cakes and gruel. In the Americas, they are still using it for this purpose (Hatfield 1971; www.cabi.org). Grain is also used as poultry and livestock feed. It may also be used as fodder for livestock (Ozturk et al. 2019).

2.10 *Chenopodium murale* L.

Family: Amaranthaceae	
Synonym(s): *Anserina muralis* (L.) Montandon; *Atriplex muralis* (L.) Crantz; *Chenopodium biforme* Nees; *Chenopodium carthagenense* Zucc.; *Chenopodium carthagenense* Zuccagni; *Chenopodium chamrium* Buch.-Ham.; *Chenopodium congestum* Hook.f.; *Chenopodium flavum* Forssk.; *Chenopodium gandhium* Buch.-Ham.; *Chenopodium guineense* Jacq.; *Chenopodium laterale* Aiton; *Chenopodium longidjawense* Peter; *Chenopodium lucidum* Gilib.; *Chenopodium maroccanum* Pau; *Chenopodium murale* f. *albescens* (Moq.) Maire; *Chenopodium murale* var. *acutidentatum* Aellen; *Chenopodium murale* var. *albescens* Moq.; *Chenopodium murale* var. *angustatum* Fenzl; *Chenopodium murale* var. *biforme* (Nees) Moq.; *Chenopodium murale* var. *carthagenense* Moq.; *Chenopodium murale* var. *latifolium* Fenzl; *Chenopodium murale* var. *microphyllum* Coss. & Germ.; *Chenopodium murale* var. *paucidentatum* Beck; *Chenopodium murale* var. *spissidentatum* Murr; *Rhagodia baccata* (Labill.) Moq.; *Rhagodia baccata* var. *congesta* (Hook. F.) Hook. F.; *Rhagodia congesta* (Hook. F.) Moq.; *Vulvaria trachisperma* Bubani (www.powo.science.kew.org).	
English name: Nettleleaf goosefoot	
Flowering period: May–July	
Distribution: Mostly in subtropics and warm-temperate regions in Asia (Malik et al. 2021).	
Altitude: Up to 2000 m	
Habit: Herb	
Ecological type: Psammohalophyte	
Habitat: Waste places, roadsides, rocks, and seashores.	
Phytochemicals: Phenols, phytosterols, glycosides, terpens, monoterpenoids, alkaloids, coumarins, Vitamins A and C, and calcium (Kokanova-Nedialkova et al. 2009).	
Potential biological activity: The plant shows antihypertensive, anthelmintic, laxative, analgesic, diuretic, and anti-inflammatory activities (Kokanova-Nedialkova et al. 2009; Ahmed et al. 2017). Many other medicinal uses such as antibacterial, antifungal, anti-diaphoretic, anti-asthmatic, migraine, digestive problems, sterility, anxiolytic, and antidepressant have been reported (Khan et al. 2019).	
Toxicity: Due to the amounts of oxalic acid and nitrate that the plant contains, a high consumption is not recommended (Guil et al. 1996, 1997).	
Botanical description: Annual, stem yellowish to green-striped, rarely tinged with red, to 70(−90) cm, usually erect, often much-branched especially in basal parts, branches ± spreading, lowermost subopposite; leaves usually olive green, sometimes yellowish or tinged with red, ± farinose; petiole shorter than or equal to the blade; blade thin to somewhat fleshy, 1–8(−10) cm long, usually broad, (triangular -) broadly ovate - broadly trullate (− rhombic), coarsely dentate, not 3-lobed, teeth often slightly incurved and acute to obtuse, apex acute to obtuse, base variously attenuate to truncate; uppermost leaves lanceolate, sometimes entire; inflorescence terminal and axillary, mostly leafy, divaricately branched, branches shorter than subtending leaves, glomerules small, ± densely arranged; flowers bisexual and sometimes female; perianth segments 5, connate below the middle, prominently keeled near apex, green, sometimes red when old, vesicular hairy; stamens 5; stigmas 2; fruits falling with perianth; pericarp prominently papillose, firmly adherent; seeds horizontal, black, 1.1–1.4 mm in diameter, round in outline, margin acutely keeled; testa prominently pitted, pits not elongated (Fig. 2.13) (www.worldfloraonline.org).	
Pollen features: Pollen type: apolar; pollen shape: spheroidal; pollen ornamentation: scabrate; polar diameter: 13.75–15.75 μm; exine thickness: 1.0–2.0 μm (Nazish et al. 2019).	
Anatomical features: Amphistomatic leaves found in the plant, with paracytic type of stomata (Fig. 2.14) (Nazish et al. 2020).	

Part used: Leaf, stem, seed, flower

Mode of utilization: Cooked, powder, paste, smoke

Route of administration: Oral, topical and inhale

Disease treated: Night blindness, snake bite, measles, hepatitis, paralysis, headache, lice killer, fever, intestinal ulcer, and obesity.

Traditional uses: The plant is used for the prevention of liver disorders; leaf decoction is used in the treatment of jaundice (Khan et al. 2019).

Recipes:
- Young shoots and leaves are cooked as vegetable often eaten to reduce obesity.
- Seed powder is taken with the powder of wheat seeds to cure hepatitis.
- Leaf paste is applied on snake bite to reduce venom effects.
- Flower paste is applied externally to cure victims of measles.
- The smoke of flowers is given before sleep to treat night blindness.

Potential economic significance: The leaves and young shoots of the plant are eaten as leafy vegetable, either steamed in its entirety, or cooked like spinach (Le Dang et al. 2010).

Fig. 2.13 *Chenopodium murale* L.

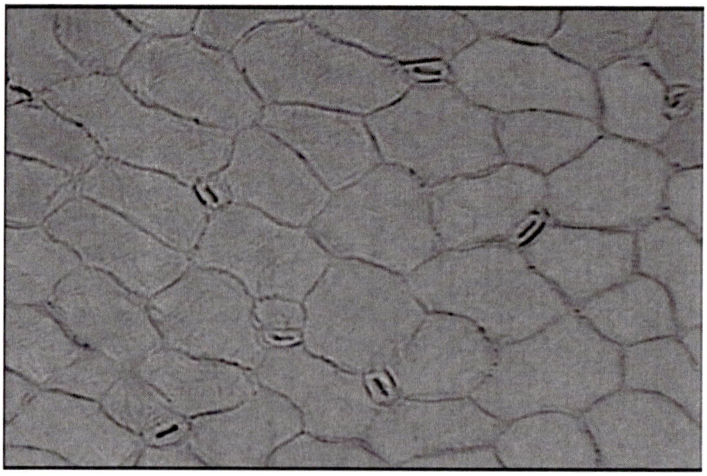

Fig. 2.14 Abaxial surface of *Chenopodium murale* L.

2.11 *Citrullus colocynthis* (L.) Schrad.

Family: Cucurbitaceae

Synonym(s): *Citrullus colocynthis* subsp. *insipidus* (Pangalo) Fursa; *Citrullus colocynthis* subsp. *stenotomus* (Pangalo) Fursa; *Citrullus colocynthoides* Pangalo; *Citrullus pseudocolocynthis* M.Roem.; *Colocynthis officinalis* Schrad.; *Colocynthis vulgaris* Schrad.; *Cucumis bipinnatifidus* Wight ex Naudin; *Cucumis colocynthis* L.; *Cucumis colocynthoides* Pi.Savi; *Cucurbita colocyntha* Link (www.powo.science.kew.org).

English name: Colocynth, Bitter Apple

Flowering period: November–July

Distribution: Afghanistan, Bangladesh, India, Iran, Iraq, Kuwait, Lebanon, Myanmar, Oman, Pakistan, Palestine, Saudi Arabia, Syria, Turkiye, Turkmenistan, and Yemen (Malik et al. 2021).

Altitude: Up to 1000 m

Habit: Herb

Ecological type: Xerophyte

Habitat: Sandy arid soils

Phytochemicals: Cucurbitacins, alkaloids, flavonoids, phenolic acids, terpenoids, glycosides, carbohydrates, fatty acids, tocophenols, carotenes, amino acids, and essential oils (Hussain et al. 2014).

Potential biological activity: Antidiabetic, anticancer, anti-inflammatory, antioxidant, antiallergic, insecticidal, and antimicrobial (Hussain et al. 2014).

Toxicity: Cytotoxic (Hussain et al. 2014; Bourhia et al. 2020).

Botanical description: A perennial herbaceous vine, with small flowers; stems angular, rough with rough hairs; leaves alternately arranged on the petioles and rough to touch, 5–10 cm in length, 1.5–2 cm in width, deeply 3–7 lobed; solitary pale yellow blooms; flowers yellow and seen on the axils of the leaves; monecious, single and pedunculated; fruits about 7–10 cm in diameter, green with undulate yellow stripes, become yellow all over when dry; the fruit bitter and globular with a smooth texture; hard and a rind around it and contains 200–300 seeds/gourd; seeds small (6 mm in length), ovoid, compressed, smooth and brownish when ripe (Fig. 2.15) (Schafferman et al. 1998; Savithramma et al. 2007; Amamou et al. 2011; Hussain et al. 2014).

Fig. 2.15 *Citrullus colocynthis* (L.) Schrad.

Pollen features: Shape: oblate-spheroidal; polar axis: 58.78 μm (47.32–71.81); equatorial diameter: 59.23 μm (53.23–64.62); exine thickness: 2.28 μm (1.79–3.59) (Perveen and Qaiser 2008).

Anatomical features: The plant has tetrarch primary root anatomy. The parenchymatous rays are relatively narrow, with parallel sides and are formed opposite to the xylem pole. The secondary xylem parts between the rays are almost fully composed of parenchyma in radial files, filled with scattered vessels or clusters of 2 to 4 vessels, each encircled by lignified parenchyma cells 1–2 layers deep (Burrows and Shaik 2015).
The leaf surface of the plant shows diacytic and cyclocytic stomata. The epidermal cell wall shape: polygonal, and epidermal cell size: 24.31–28.69 μm (Abdulrahaman et al. 2011).

Part used: Fruit, root, seed

Mode of utilization: Decoction, paste, extract, raw, powder

Route of administration: Oral, topical

Disease treated: Constipation, gastric problems, hemorrhoids, dysentery, liver diseases, excessive menstruation, legs tingling, diabetes, digestive disorders, sciatica, gall bladder stones, toothache, sexual tonic, respiratory diseases, and blood purification.

Traditional uses: In Pakistan, India, and China, traditionally the plant is used for gut disorders, indigestion, dysentery, gastroenteritis, constipation, leprosy, asthma, bronchitis, jaundice, joint pain, cancer and mastitis and colic pain as well as common cold, cough, toothache, wounds, and diabetes (Qureshi et al. 2010; Amamou et al. 2011; Gurudeeban et al. 2011; Hussain et al. 2014; Malik et al. 2021).

Recipes:
- Fresh fruit pieces are boiled in water with sugar to make paste which is used for constipation.
- Root decoction is used as ear drops to get relieve from earache.
- Fruit paste is recommended to treat diabetes and to improve eyesight.
- Tooth stick made from root is used to relieve pain from toothache.
- Fruit decoction is taken orally to treat sciatica.
- Dried unripe fruit powder is taken with milk for gall bladder stones.
- Half spoon of root extract is taken orally before breakfast to cure digestive disorders.
- Seed flour is used to make loaf which is consumed to cure diabetes.

Potential economic significance: The fruits of the plant are used as food for animals and humans (Sadou et al. 2007). The nutritional and functional features of seeds have been studied at length by the United States Department of Agriculture (USDA). They have concluded that this plant has potential to find a place in the food industry (National Research Council 2006; Hussain et al. 2014).

2.12 *Cleome viscosa* L.

Family: Cleomaceae
Synonym(s): *Arivela viscosa* (L.) Raf.; *Arivela viscosa* var. *deglabrata* (Backer) M.L.Zhang & G.C.Tucker; *Cleome acutifolia* Elmer; *Cleome icosandra* L.; *Cleome viscosa* var. *deglabrata* (Backer) B.S.Sun; *Cleome viscosa* f. *deglabrata* (Backer) Jacobs; *Cleome viscosa* var. *nagarjunakondensis* Sundararagh.; *Cleome viscosa* var. *parviflora* Kuntze; *Cleome viscosa* var. *viscosa*; *Polanisia icosandra* (L.) Wight & Arn.; *Polanisia microphylla* Eichler; *Polanisia orthocarpa* Hochst. ex Webb; *Polanisia viscosa* (L.) Blume; *Polanisia viscosa* var. *deglabrata* Backer; *Polanisia viscosa* var. *icosandra* (L.) Schweinf. ex Oliv.; *Sinapistrum viscosum* (L.) Moench (www.powo.science.kew.org).
English name: Asian spider flower
Flowering period: June–September
Distribution: South-Eastern Asia, Southern Arabia, and throughout India (Das and Layek 2021).
Altitude: Up to 1000 m
Habit: Herb
Ecological type: Xerophyte
Habitat: The plant grows under warm and wet conditions on sandy soils, but sometimes on calcareous and rocky soils (Graveson 2012).
Phytochemicals: Flavonoids, glycoflavanone, diterpene lactones, triterpenes, sterol, coumarins, glycosides, saponin, fatty acids, amino acids, and sucrose (Mali 2010; Bose et al. 2011; Das and Layek 2021).
Potential biological activity: The plant possesses antipyretic, antidiarrheal, anti-inflammatory, antimicrobial, immunomodulatory, local anesthetic, and antimalarial activities (Mali 2010; Bose et al. 2011).
Toxicity: Cytotoxic (Bose et al. 2011).
Botanical description: Erect yellowish-green annual herb; stem densely clothed with glandular and simple hairs, grooved; leaves 3–5 foliate, leaflets elliptic-oblong, entire, subacute, central one largest, all subsessile; flowers yellow, axillary, growing out into a lax raceme, sepals oblong-lanceolate, glandular, hairy outside, stamens 12–20; fruits capsule linear, erect, narrowed at both ends, viscid, pubescent, striate (Fig. 2.16) (Das and Layek 2021).
Pollen features: Pollen grains prolate, tricolporate, monad; polar axis 19.5 μm and equatorial axis 18.9 μm; exine 2.6 μm (Kasem 2016).
Anatomical features: Onoja (2016) has reported that the lower (abaxial) and upper epidermis (adaxial) contain anomocytic stomata and glandular trichomes as the distinguishing features. The epidermis has a single-layered round cell, hypodermis has two layers, the vascular bundles are not conjoint and surrounded by bundle sheath (Kasem 2016). Epidermis cells are irregular in shape and cell walls are sinuous in outline. Leaves are amphistomatic. Ca-oxalate crystals are found on both the epidermal layers (Das and Layek 2021).
Part used: Seed, leaf, root, and stem
Mode of utilization: Powder, decoction, paste, oil, and cooked
Route of administration: Oral, topical
Disease treated: Ring worms, improve appetite, digestive disorders, diarrhea, increase lactation, wound healing, rheumatism, urethral ulcer, earache, sexual tonic, bone fractures, abortion, insect repellent, gastric pain, premature ejaculation, and hiccups.

Fig. 2.16 *Cleome viscosa* L.

Traditional uses: The plant is used in uterine complaints, leprosy, treatment of malarial fevers and those due to indigestion, skin / blood diseases, as laxative and diuretic (Kirtikar and Basu 1987; Mali 2010; Bose et al. 2011). The seeds are also documented as anthelmintic and given to treat fever and diarrhea in the Unani system of medicine (Chopra et al. 1958; Bose et al. 2011). According to Bose et al. (2011), a poultice made from the plant is efficacious as a counter irritant in chronic painful joints and the juice is used to remove pus from the ear.

Recipes:

- Seed powder is given with milk to treat ring worms.
- Boiled leaves are eaten to improve appetite.
- Leaf decoction is digestive stimulant and cure diarrhea.
- Leaf powder taken with water increase lactation in pregnant women.
- Seed decoction is used externally to cure rheumatism.
- Root paste is used externally to cure earache.
- Seed oil is used externally to treat urethral ulcer.
- Young crushed stem is cooked with salt and chilies and eaten as sexual tonic.

Potential economic significance: The pleasant flavoring seeds are used in India in the preparation of vegetables, curries, sausages, pulses and pickling spices as a condiment in place for mustard seed and cumin The narcotic properties of tobacco are reported to increase if dried and powdered leaves and seeds are added to it as reported from Sumatra (Windadri 2001).

2.13 *Cressa cretica* L.

Family: Convolvulaceae	
Synonym(s): *Cressa ballii* Batt.; *Cressa humifusa* Lam.; *Cressa indica* Retz.; *Cressa intermedia* T.Anderson; *Cressa microphylla* St.-Lag.; *Cressa monosperma* Stokes; *Cressa villosa* Hoffmanns. & Link (www.powo.science.kew.org).	
English name: Alkali weed, Rosin weed, Cressa	
Flowering period: June–August	
Distribution: Afghanistan, Bangladesh, India, Iran, Iraq, Kuwait, Lebanon, North Caucasus, Oman, Pakistan, Palestine, Saudi Arabia, Syria, Tadzhikistan, Transcaucasus, Turkiye, Turkmenistan, Uzbekistan, Yemen (Malik et al. 2021).	
Altitude: sea level −175 m	
Habit: Herb	
Ecological type: Hydrohalophyte	
Habitat: On sandy or muddy saline habitats along the seacoast (Priyashree et al. 2010).	
Phytochemicals: Alkoloids, flavonoids, terpenes, tannins, saponins, phytosterols, coumarins, glycosides, fatty acids, and vitamin E (Rani et al. 2011; Al-Snafi 2016).	
Potential biological activity: Antibacterial, antifungal, antidiabetic, antitussive, antioxidant, anti-inflammatory, antipyretic, analgesic, testicular functions and antifertility activities (Priyashree et al. 2010; Rani et al. 2011; Al-Snafi 2016).	
Toxicity: Cytotoxicity and genotoxicity (Mutlag et al. 2017; Mahdi and Abaas 2019).	
Botanical description: Stems basally woody, much branched above, branches pilose; leaves closely condensed, sessile, ovate-lanceolate to ovate, 2.5–9 × 1–6 mm, apex acute, basally cuneate, rounded or sub-cordate, pubescent; calyx ovoid, 3–4 mm long, the sepals obovate, abruptly acute, silky pubescent; corolla 5–6 mm long, the tube and the oblong-ovate lobes about equal in length; stamens and styles exerted for about 2–4 mm; ovary 2-celled, 4-ovuled; fruit ovoid, 3–4 mm long; seed usually 1, shining (Fig. 2.17) (www.efloras.org).	

Fig. 2.17 *Cressa cretica* L.

Pollen features: pollen class: tricolpate; polar axis: 30–33 μm; equatorial diameter: 35–37 μm; pollen shape: suboblate (Osman and Abdel Khalik 2005).

Anatomical features: According to Rajput et al. (2014), the young stems are thick, circular in outline, with stomata and non-glandular multiseriate trichomes in the epidermis, which is single-layered and composed of thin-walled square parenchyma cells covered by a thin cuticle in young stems. The epidermal cells are replaced by cork cells in the thick stems. The cortex is fairly wide in the young stems, composed of small and large cells. The outer part of cortex has irregularly distributed, thick-walled, lignified sclerenchymatous fibers. The large cells are oval to polygonal, arranged in radial files; smaller cells are oval to circular, placed at the corners of large cells. The endodermis is composed of single layer of barrel-shaped cells formed on the inner side of cortex, with pericycle lying beneath it. The collateral vascular bundles 8–10 are joined by interfascicular cambium forming a complete ring of vascular cambium (Rajput et al. 2014).

Part used: Whole plant

Mode of utilization: Decoction, paste

Route of administration: Oral, topical

Disease treated: Asthma, expectorant, and skin diseases

Traditional uses: The plant is used in diabetes, asthma, expectorant, stomachic, and antibilious; also used as anthelmintic, stomachic, tonic and aphrodisiac, enrich the blood and is useful in constipation, leprosy, asthma and urinary discharge (Rani et al. 2011).

Recipes:

- A decoction of the whole plant is drunk twice a day as a remedy for asthma and expectorant (Qasim et al. 2014).
- A paste of the whole plant is applied locally in skin diseases especially boils (Katewa et al. 2004).

Potential economic significance: The fruits are a potential source of edible oil (Priyashree et al. 2010). This halophyte plant is considered as a suitable candidate for biodiesel production due to high-level seed oil contents besides good quality engine parameters (Jaafar et al. 2021). The plant is also used for camel, goat, and sheep feed fodder (Ozturk et al. 2019).

2.14 *Cyperus rotundus* L.

Family: Cyperaceae	
Synonym(s): *Chlorocyperus rotundus* (L.) Palla; *Chlorocyperus salaamensis* Palla; *Cyperus agrestis* Willd. ex Spreng. & Link; *Cyperus arabicus* Ehrenb. ex Boeckeler; *Cyperus badius* var. *inconspicuus* (Nyman) Nyman; *Cyperus bicolor* Vahl; *Cyperus bifax* C.B.Clarke; *Cyperus bulbosostoloniferus* Miq.; *Cyperus comosus* Sm.; *Cyperus disruptus* C.B.Clarke; *Cyperus elongatus* Sieber ex Kunth; *Cyperus herbicavus* Melliss; *Cyperus hexastachyos* Rottb.; *Cyperus hildra* Poir.; *Cyperus hydra* Michx.; *Cyperus inconspicuus* Gennari; *Cyperus laevissimus* Steud.; *Cyperus leptostachyus* Griff.; *Cyperus longus* Boeckeler; *Cyperus merkeri* C.B.Clarke; *Cyperus micreilema* Steud.; *Cyperus nubicus* C.B.Clarke; *Cyperus ochreoides* Steud.; *Cyperus officinalis* Nees ex Godr.; *Cyperus oliganthus* Gand.; *Cyperus olivaris* O.Targ.Tozz.; *Cyperus pallescens* Poir.; *Cyperus patulus* M.Bieb.; *Cyperus platystachys* Cherm.; *Cyperus procerulus* Nees; *Cyperus proteinolepis* Boeckeler; *Cyperus pseudovariegatus* Boeckeler; *Cyperus purpureovariegatus* Boeckeler; *Cyperus radicosus* Sm.; *Cyperus retzii* Nees; *Cyperus rubicundus* Willd. ex Link; *Cyperus rudioi* Boeckeler; *Cyperus stoloniferus* var. *pallidus* Boeckeler; *Cyperus taylorii* C.B.Clarke; *Cyperus tenuifolius* Walp.; *Cyperus tetrastachyos* Desf.; *Cyperus tuberosus* Rottb.; *Cyperus viridis* Roxb. ex C.B.Clarke; *Cyperus weinlandii* Kük.; *Cyperus yoshinagae* Ohwi; *Pycreus rotundus* (L.) Hayek; *Schoenus tuberosus* Burm.f. (Malik et al. 2021).	
English name: Nutgrass, Nutsedge	
Flowering period: April–October	
Distribution: Afghanistan, Bangladesh, Borneo, Cambodia, China, East Himalaya, India, Iran, Iraq, Kazakhstan, Kyrgyzstan, Lebanon, Myanmar, Nepal, North Caucasus, Oman, Pakistan, Palestine, Philippines, Saudi Arabia, Sri Lanka, Syria, Taiwan, Thailand, Tibet, Transcaucasus, Turkiye, Turkmenistan, Uzbekistan, Vietnam, West Himalayas, and Yemen (Malik et al. 2021).	
Altitude: Up to 1000 m	
Habit: Herb	
Ecological types: Hydrohalophyte	
Habitat: Agricultural areas, coastland, riparian zones, and watercourses (Malik et al. 2021).	
Phytochemicals: Polyphenolic, alkaloids, cyperols, flavonoids, fatty oil, furochromone, glycerols, linolenic acids, myristic acids, nootkatones, saponin, sesquiterpene, sitosterols, stearic acids, terpenoid, valencenes, steroid glycoside, sesquiterpene hydrocarbon, α-cubebene, and cyprotene (Malik et al. 2021).	
Potential biological activity: Antidepressant, anti-hepatitis B, antimalarial, antidiarrheal and antioxidant activities (Al-Massarani et al. 2016).	
Toxicity: Fumigant toxicity (Chang et al. 2012), and phytotoxicity (El-Rokiek et al. 2010).	
Botanical description: Perennial, 21–61 cm long; rhizomes dark- brownish, woody; stolons several, long-creeping, stout, 2–1.6 mm wide; stems 1.0–2.6 mm wide, trigonous, greenish, soft; leaves shorter than stems; sheath 11 cm long, wider, greyish brown, acute, or brownish, sometimes red-tinted, with lingulate margin; blade 31 cm long, 1.0–4.6 mm in diameter, greyish green, folded, with a slight keel, margin soft towards the barbed long apex, trigonous, acute, scabrous; inflorescences an antheridium, 41–101 mm long; 3–4 lower bracts longer than inflorescences; spike 11–41 × 1.6–2.1 mm long, with 12–52 glumes, glume-like bracts 1.8–2.5 mm up to 3 mm long, spikes 3-angle, 0.6 mm broad, internode 0.8–1.2 mm in length, winged; glumes (2.4-)2.6–3.6 mm, nerve areas wider, margins narrow sometimes wide scarious, embracing towards the apex; nuts 1.4–1.7 × 0.6–1.1 mm length, oval, trigonous, brownish to black brownish, obscure, reticulated papillose (Malik et al. 2021) (Fig. 2.18).	
Pollen features: Pollen shape: sub oblate, polar length: 25–31 µm, equatorial length: 27–40 µm, ornamentation surface: granulate (Sosam and Al-Mayyahi 2018).	

Fig. 2.18 *Cyperus rotundus* L.

Anatomical features: A single layer of large epidermal cells forms the upper leaf surface with a thick cuticle covering it (Wills and Briscoe 1970). The stomata are distributed only on the lower surface; vascular bundles vary from collateral to amphivasal; passing from the leaves through the bulb into the rhizomes and tubers. The rhizomes and tubers which have developed newly are white and fleshy with a parenchymatous epidermis and cortex. We find brown and wiry mature rhizomes with a deteriorated outer cortex, a lignified inner cortext and endodermis. The rhizome apices are beset tubers and bulbs, each accumulating starch. The interconnecting vascular system appears to remain intact throughout the growing season (Wills and Briscoe 1970).

Part used: Flowers, whole plant

Mode of utilization: Powder, infusion

Route of administration: Oral and topical

Disease treated: Skin diseases

Traditional uses: Several researchers have reported (Zhu et al. 1997; Joshi and Joshi 2000; Puratuchikody et al. 2006; Sivapalan 2013) its evaluation as antimalarial agent, for stomach disorders, wounds, boils, as an analgesic, sedative, antispasmodic, in blisters and to relieve diarrhea. According to Yu et al. (2004) and Nima et al. (2008), this plant tuber is one of the oldest known medicines used for the treatment of dysmenorrheal and menstrual irregularities. Herbal infusion has been used in pain, fever, diarrhea, dysentery, an emmenagogue, and other intestinal problems (Umerie and Ezeuzo 2000; Malik et al. 2021).

Historically, the tubers of this species were roasted in the Arabian countries and hot ashes from burned tubers were used to treat carbuncles, bruises, wounds, and other related complaints (Al-Massarani et al. 2016). In the Islamic herbal medicine and Ayurvedic practices the plant tubers are reported to have been used to treat delayed menstruation and dysmenorrhea, fever, uterine disorders, as stomachic, removal of obstructions and as emollient plaster (Pirzada et al. 2015; Al-Massarani et al. 2016).

In the traditional Chinese medicine, the tubers of the plant are used as an antidiarrheal, antidepressant, analgesic, antiinflammatory and antiemetic remedy for dysentery and women's diseases (Chen et al. 2014; Oh et al. 2015; Al-Massarani et al. 2016).

Recipes:

- For 20 min 50 g of dry flower crushed into powder, 2 tablespoons used with water daily in the evening for treating boils (Malik et al. 2021).
- For 3 days 50 g of whole plant infused in 20 mL of water; infusion filtered and 2 cups (200 mL) of infusion taken in the morning for 7 days daily for treating skin diseases (Malik et al. 2021).

Potential economic signifance: The plant is harvested from the wild for local use and trade. It is used mainly as a medicine but also for food, essential oils and basketry; sometimes also cultivated in tropical regions for its edible tubers (www.tropical.theferns.info).

2.15 *Dactyloctenium scindicum* Boiss.

Family: Poaceae	
Synonym(s): *Dactyloctenium glaucophyllum* Courbon; *Eleusine aristata* Ehrenb. ex Boiss.; *Eleusine glaucophylla* (Courbon) Munro ex Benth.; *Eleusine scindica* (Boiss.) Duthie (www.powo.science.kew.org).	
English name: Crowfoot grasses	
Flowering period: July–September	
Distribution: Afghanistan, India, Oman, Pakistan, Saudi Arabia, West Himalayas, Yemen (Malik et al. 2021).	
Altitude: Up to 1400 m	
Habit: Herb	
Ecological type: Xerohalophyte	
Habitat: Dry rocky outcrops, sandy soil on hard ground and on wet ground; also grow on highly saline soil tracks (Landge et al. 2021).	
Phytochemicals: Not reported	
Potential biological activity: Not reported	
Toxicity: Not reported	
Botanical description: Stoloniferous perennial forming extensive spreading mats; culms slender with swollen bases, 7–45 cm high, erect; leaf-blades flat or loosely folded, tough and rather glaucous, 1–11 cm long, 1.5–3 mm wide, scattered papillose-hispid especially along the margins; inflorescence composed of 34(5) slightly falcate oblong spikes 0.8–2 cm long forming a compact head, the spikes readily disarticulating at maturity from the top of the culm; spikelets 3–9-flowered, broadly lanceolate to ovate, 4–8 mm long; glumes broadly elliptic in profile, the lower 1.7–2.5 mm long, the keel narrowly winged, the upper 1.5–2.3 mm long, the prominent scabrid keel extended into an awn half to as long as the glume; lemmas lanceolate in profile, 3–3.8 mm long, obtuse to acute with the scabrid keel extended into a mucro up to 0.8 mm long; palea-keels finely scabrid, unwinged; anthers 1.1–2 mm long; Grain 0.7–1 mm long, transversely rugose (Fig. 2.19) (www.efloras.org).	
Pollen features: Not reported	

Fig. 2.19 *Dactyloctenium scindicum* Boiss.

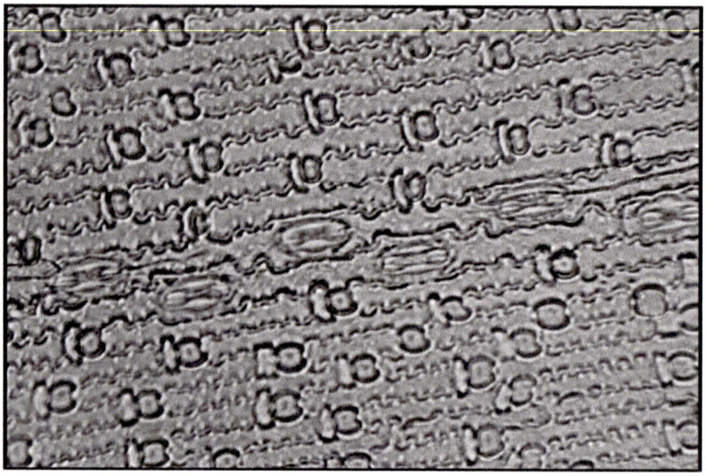

Fig. 2.20 Abaxial surface of *Dactyloctenium scindicum* Boiss.

Anatomical features: Lamina transverse sections depict slight ribs and furrows on the adaxial but flat on the abaxial surface; the vascular bundles are mostly with adaxial sclerenchyma strands; the chlorenchyma cells are radially arranged around the vascular bundles, the keel is conspicuous, rounded with single median vascular bundle, having 2–3 small vascular bundles on each side; the bulliform and associated colorless cells are in the form of fan-shaped groups; the large vascular bundles possess double sheath, complete or interrupted on the adaxial side, due to colorless cells; in the small vascular bundles inner sheath is not prominent; the stomatal complex in general has guard cells dumbbell-shaped, subsidiary cells triangular or high dome-shaped, together with micro -macro hairs; hooks are absent (Fig. 2.20) (Ahmad et al. 2011).

Part used: Leaf, stem, root, seed, flower, seed

Mode of utilization: Paste, extract, powder, juice, paste, decoction

Route of administration: Oral and topical

Disease treated: Syphilis, liver diseases, leukoderma, spleen enlargement, abscess, hydrophobia, toothache, gonorrhea, ulcer, earache, snake bite, analgesic, heart burn, kidney diseases, diuretic, and wounds.

Traditional uses: *Pennisetum typhoides* seeds are mixed with the seed powder of this species and used in making bread (roti); entire plant extract is used orally in rheumatism (Katewa and Galav 2006; Ahmad et al. 2020). In addition, the plant is used to treatment of pain killer, inflammation, and arthritis (Qasim 2014).

Recipes:

- Leaf paste is applied dermally to treat itching of wounds.
- Culm extract after strained with milk is taken orally to cure gonorrhea.
- Root powder is taken orally to cure ulcer.
- The juice of leaves is used as ear drops to get relief from earache.
- Root extract is applied on snake bite dermally to reduce the venom effect.
- Leaf paste is taken with milk to get relief from heartburn.

Potential economic significance: It is one of the well consumed plants by animals and supposed to increase milk yield (especially cattle) (Singh et al. 2012; Gul et al. 2014).

2.16 *Halocnemum strobilaceum* (Pall.) M.Bieb.

Family: Amaranthaceae

Synonym(s): *Halocnemum cruciatum* (Forssk.) Baill.; *Halocnemum strobilaceum* subsp. *Cruciatum* (Forssk.) Arrigoni; *Salicornia fruticosa* Forssk.; *Salicornia drepanensis* Tineo ex Ung.-Sternb.; *Salicornia fruticosa* Moq.; *Salicornia glauca* Sieber ex Ung.-Sternb.; *Salicornia strobilacea* Pall.; *Sarcathria strobilacea* Raf. (www.powo.science.kew.org).

English name: Halocnemum

Flowering period: August–October

Distribution: Afghanistan, China, Iran, Iraq, Kazakhstan, Kyrgyzstan, Kuwait, Lebanon, Mongolia, North Caucasus, Oman, Pakistan, Palestine, Russia, Saudi Arabia, Syria, Tadzhikistan, Transcaucasus, Turkiye, Turkmenistan, Uzbekistan, West Siberia (Malik et al. 2021).

Altitude: Up to 1200 m

Habit: Shrub

Ecological type: Xerohalophyte

Habitat: Salt-lake shores, and salt marshes

Phytochemicals: Saponins, flavonoids, alkaloids, anthocyanins, coumarins, and fatty acids (Nasernakhaei and Zahraei 2021).

Potential biological activity: Anticancer, antioxidant, anti-inflammation, and antimicrobial activities (Ksouri et al. 2012; Nasernakhaei and Zahraei 2021).

Toxicity: Cytotoxicity (Razek et al. 2019; Pourabdollah Kaleybar 2021).

Botanical description: Plants 20–40 cm tall; stems branched from base; old branches nearly alternate, prostrate or ascending, brown-green, woody, bearing opposite, shortened, budlike, dwarf branchlets; young branchlets opposite, suberect, gray-green, jointed, glabrous; leaves opposite, connate; spikes borne on upper branches, decussate, 5–15 × 2–3 mm; perianth with 2 lateral segments incurved, outline obdeltoid; seed brown, ovoid or globose, 0.5–0.75 mm in diam., densely finely papillate (Fig. 2.21) (www.efloras.org).

Fig. 2.21 *Halocnemum strobilaceum* (Pall.) M.Bieb.

Pollen features: Pollen diameter: 20.77 μm; exine thickness: 2.61 μm (Dehghani et al. 2021).

Anatomical features: According to Yanbin et al. (1998), the leaves and assimilating branches of this species are adapted to salt-morphic circumstances due to the possession of thick leaf blades and cuticula; epidermal hairs; sunken stomata, well developed palisade and many equilateral leaves. The assimilating branches take place of the photosynthetic function of some modified leaves; leaves and assimilating branches possess mucilage or crystal cells; well-developed water-storing tissues with large cells are observed in the interior of the assimilating branches (Yanbin et al. 1998).

Part used: Aerial parts

Mode of utilization: Herbal tea, brew, decoction, extract

Route of administration: Oral

Disease treated: Pregnancy difficulties, dysmenorrhea, headache, expectorant, and digestive diseases.

Traditional uses: The plant is used to treat fever, jaundice, headache, expectorants, digestion, pregnancy difficulties, and dysmenorrhea, and hair loss (Nasernakhaei and Zahraei 2021).

Recipes:

- The plant extract prepared with herbal tea is used to treat of pregnancy difficulties and dysmenorrhea (Heneidy et al. 2017).
- A decoction and/or brew of the aerial parts of the plant is drunk as treatment of headache, expectorant, and digestive remedy (Zali and Tahmasb 2016).

Potential economic significance: The aerial parts of the plant are used to get extract to be used as a probiotic in aquaculture feed (Messina et al. 2019). The plant is a good source of vitamins C and E as well as sodium, potassium, calcium, and magnesium. The green or red-violet shoots are consumed in Tunisia as gourmet vegetable due to their organoleptic properties and are used as preservatives in the food and beverage industries (Zaier et al. 2020).

2.17 *Juncus acutus* L.

Family: Juncaceae
Synonym(s): *Juncus acutus* subsp. *acutus*; *Juncus acutus* subsp. *leopoldii* (Parl.) Snogerup
English name: Spiny rush
Flowering period: March–June
Distribution: Iran, Iraq, Kuwait, Jordan, Lebanon, North Caucasus, Palestine, Syria, Transcaucasus, Turkiye, and Turkmenistan
Altitude: Up to 150 m
Habit: Herb
Ecological type: Hydrohalophyte
Habitat: In sandy seashores, salt marshes, and disturbed saline areas, freshwater wetlands, and saline and subsaline wetlands.
Phytochemicals: Phenolics, flavonoids, phenanthrenoids, coumarines, carotenoids, and juncunol (Petropoulos et al. 2018).
Potential biological activity: Antioxidant, anti-inflammatory, and anti-acetylcholinesterase (Petropoulos et al. 2018).
Toxicity: Cytotoxic (Rodrigues et al. 2014).
Botanical description: Plant densely cespitose with poorly developed rhizome; intravaginal shoots present; stems (15-)50–150 cm, rigid, with 2–5 basal leaves; inflorescence many-flowered, usually dense, globose; lowest 2 bracts well developed although usually short; perianth segments subequal, oblong; outer obtuse or acute; inner with broad scarious apical auricles; stamens equaling or slightly shorter than perianth; anthers 1.5–2 mm, several times longer than filaments; capsule ovoid, 4–6 mm, much-exceeding perianth, with broadly conical apex, shortly mucronate; seeds 80–120 per capsule, 1.5–2.5 mm, with equal appendages (Fig. 2.22) (Davis 1985).
Pollen features: Not reported

Fig. 2.22 *Juncus acutus* L.

Anatomical features: The anatomical structure of roots of the plant reveals these are adventitious and very thin, the rhizodermis has relatively short, abundant absorbing hairs (Al Hassan et al. 2015). The cortex is relatively thick with an exodermis having 2–3 layers of cells having thin, less suberized walls; cortical parenchyma is thick, with many large air-storing cavities, very close to each other and separated only by fragments of disintegrated parenchymatous cells plus stele; the endodermis has large cells, with internal and radial walls moderately thickened and lignified. The stele is relatively thick, with a pericycle having smaller cells compared to the endodermis, and alternating with these, all walls are thin, in moderately sclerified and lignified parenchyma, 7–8 large metaxylem vessels are embedded in a circle (Al Hassan et al. 2015).
Part used: Seed
Mode of utilization: Eaten fresh
Route of administration: Oral
Disease treated: Diuretic
Traditional uses: Urinary diseases, jaundice (Miara et al. 2021).
Recipes: The seeds of the plant employed as diuretic (La Rosa et al. 2021).
Potential economic significance: Generally, the stems of the plant are used in making woven baskets, thatching, and weaving mats.

2.18 *Limonium sinuatum* (L.) Mill.

Family: Plumbaginaceae	
Synonym(s): *Linczevskia sinuata* (L.) Tzvelev; *Statice sinuata* L.; *Taxanthema sinuatum* (L.) Sweet	
English name: Sea lavender, sea pink	
Flowering period: May–September	
Distribution: Lebanon, Palestine, Syria, Transcaucasus, and Turkiye	
Altitude: Up to 300 m	
Habit: Herb	
Ecological type: Hydrohalophyte	
Habitat: Coastal areas	
Phytochemicals: Phenolic acids, flavonoids (Petropoulos et al. 2018).	
Potential biological activity: Antioxidant activity (Xu et al. 2017).	
Toxicity: Heavy metal toxicity (Ansari et al. 2018).	
Botanical description: Annual or perennial herb, densely covered throughout with spreading fairly long stiff hairs; leaves radical, oblong-lanceolate, 3–10 cm long, 1–3 cm broad, deeply sinuate-margined to pinnatifid or pinnatisect; terminal lobe piliferous; scape dichotomously branched in the upper part or nearly at the top, broadly winged, produced at the nodes into 2–3 linear foliaceous appendages; spikes compact; spikelets 3–4-flowered; outer bract narrowly triangular, 6–9 mm long, membranous, subulate acuminate; first inner bract slightly longer and much broader than the outer bract, herbaceous, strongly convex with two ciliate dorsal heads, and produced into 2–3 large, 2–3-toothed appendages; other inner bracts like the outer bract; calyx funnel-shaped; tube puberulous; limb truncate, violet or yellow; nerves terminating below the margin (Fig. 2.23) (www.efloras.org).	

Fig. 2.23 *Limonium sinuatum* (L.) Mill.

Pollen features: Pollen unit: monad; size (pollen unit): 51–100 μm; pollen class: colpate (www.paldat.org).

Anatomical features: The leaves of the plant are amphistomatic type (de Blas 1992).

Part used: Flower, leaves

Mode of utilization: Decoction

Route of administration: Oral

Disease treated: Anti-aging properties

Traditional uses: This plant is used as a food plant in Cyprus (Della et al. 2006).

Recipes:

- A decoction of the flowers of the plant is used for their anti-aging properties (Xu et al. 2017).
- A decoction of the leaves of the plant is used as treatment for diabetes (Idm'hand et al. 2020).

Potential economic significance: The plant is an economically valuable cut and dried flower crop (Grieve et al. 2005).

2.19 *Noaea mucronata* (Forssk.) Asch. & Schweinf.

Family: Amaranthaceae	
Synonym(s): *Anabasis echinus* M.Bieb.; *Anabasis spinosissima* L.f.; *Halogeton spinosissimus* (L.f.) C.A.Mey.; *Noaea spinosissima* (L.f.) Moq.; *Salsola camphorosmoides* Desf.; *Salsola mucronata* Forssk. (www.powo.science.kew.org).	
English name: Thorny saltwort	
Flowering period: August–October	
Distribution: Afghanistan, Iran, Iraq, Lebanon, North Caucasus, Palestine, Saudi Arabia, Syria, Transcaucasus, Turkiye, Turkmenistan, and Uzbekistan (Malik et al. 2021).	
Altitude: 500–2200 m	
Habit: Shrub	
Ecological type: Xerohalophyte	
Habitat: Arid and semi-arid steppes and deserts (www.cabi.org).	
Phytochemicals: Alkaloids, flavonoids, saponins, amino acids, and fatty acids (Aynehchi et al. 1981).	
Potential biological activity: Anti-inflammatory, analgesic, and anti-diabetic activities (El Eraky 2001).	
Toxicity: Heavy metal toxicity (Rizwan et al. 2014).	
Botanical description: The plant is a low, much-branched shrub, 20–50 (−75) cm high, with stems hardened at their base; branches rigid, spine-tipped although with no stipules; leaves 0.5–1.0 (−1.5) cm long, glabrous, alternate and very narrow, cylindrical or terete, filiform and mucronate; flowers green, hermaphrodite, solitary and axillary, situated at the axils of the leaves; perianth segments 5 (3 outer and 2 inners), around 4 mm long, all developing a transverse wing on the back in the white-reddish fruit with wings 3–6 mm, obovate or obovate- to circular with an irregularly toothed margin; stamens 5, stigmas 2; seeds vertical (Fig. 2.24) (www.cabi.org).	

Fig. 2.24 *Noaea mucronata* (Forssk.) Asch. & Schweinf.

Pollen features: Pollen grains: pantoporate; pollen grains: 12.2–15.6 μm in diameter (Hayrapetyan and Sonyan 2021).

Anatomical features: Zarinkamar (2006) reported that various unicellular, simple trichomes, extremely short sometimes similar to papillae accompanied by glandular trichomes are present on both leaf surfaces. In addition, the epidermis includes small polygonal cells with thick smooth walls. A stoma is brachyparacytic some anomocytic is present too. Numerous large crystals (druses) are observed at the leaf margin (Zarinkamar 2006).

Part used: Leaves, stem

Mode of utilization: Infusion, decoction

Route of administration: Oral

Disease treated: Carminative, sudorific, antispasmodic, and colic.

Traditional uses: In Iran, the plant is used as a means of breaking up kidney stones (Ghasemi Pirbalouti et al. 2013).

Recipes:

- An infusion of leaf is used as a carminative, sudorific, and antispasmodic (Al-Khalil 1995).
- Young stem decoction is used to treat colic in acute cases (Al-Khalil 1995).

Potential economic significance: The plant is commonly used as fuel for cooking and heating (Al-Oudat et al. 2005). The soft shoots are occasionally consumed in spring by sheep and goats in west Asia (especially in Syria) (Al-Oudat et al. 2005; Ozturk et al. 2019). It is an effective accumulator, worth evaluation in heavy metal phytoremediation of polluted sites; good accumulator of lead, zinc, copper, cadmium out of the 5 plant species tested in Iran (Chehregani et al. 2009).

2.20 *Pentanema britannica* (L.) D.Gut.Larr., Santos-Vicente, Anderb., E.Rico & M.M.Mart.Ort.

Family: Asteraceae

Synonym(s): *Aster britannicus* (L.) All.; *Aster orientalis* S.G.Gmel.; *Aster undulatus* Moench; *Conyza britannica* (L.) Moris ex Rupr.; *Helenium britannica* (L.) Kuntze; *Helenium macrolepis* Kuntze; *Helenium microcephalum* Kuntze; *Helenium repandum* Kuntze; *Inula britannica* L.; *Inula britannica* var. *angustifolia* Becker; *Inula britannica* var. *eglandulosa* R.Nabiev; *Inula britannica* subsp. *hispanica* (Pau) O.Bolòs & Vigo; *Inula britannica* subsp. *latifolia* U.P.Pratov & R.Nabiev; *Inula britannica* var. *longilepis* R.Nabiev; *Inula britannica* var. *ramosissima* Ledeb.; *Inula britannica* var. *sericans* Zalewski; *Inula britannica* f. *sublanata* (Kom.) Kitag.; *Inula britannica* var. *sublanata* Kom.; *Inula britannica* var. *tymiensis* Kudô; *Inula chinensis* (Kom.) Kom.; *Inula comosa* Lam.; *Inula dichotoma* Zuccagni; *Inula encelioides* Hornem. ex Ledeb.; *Inula hirta* Pollich; *Inula micranthos* DC.; *Inula microcephala* Borbás; *Inula oetteliana* Rchb.; *Inula orientalis* d'Urv. ex Boiss.; *Inula serrata* Gilib.; *Inula squarrosa* Krock.; *Inula tymensis* Kudô; *Inula vaillantii* Schur ex Nyman (www.powo.science.kew.org).

English name: Meadow fleabane, yellow starwort

Flowering period: May–September (Malik et al. 2021).

Distribution: China, Iran, Iraq, Kazakhstan, Kyrgyzstan, Mongolia, Nepal, North Caucasus, Russia, Tadzhikistan, Tibet, Transcaucasus, Turkiye, Turkmenistan, Uzbekistan, West Himalayas, West Siberia (Malik et al. 2021).

Altitude: Up to 2500

Habit: Herb

Ecological type: Hydrohalophyte

Habitat: In semi-desert areas, steppes, and meadows, along the edges of rivers and lakes (Zaurov et al. 2013; Malik et al. 2021).

Phytochemicals: Alkoloids, terpenoids, flavonoids, tannins, essential oils and vitamin C (Khan et al. 2010; Zaurov et al. 2013).

Potential biological activity: Anticancer, antibacterial, antifungal, antioxidant, anti-inflammatory, neuroprotective and hepatoprotective activities (Khan et al. 2010; WHO 2013).

Toxicity: Cytotoxicity (Khan et al. 2010).

Botanical description: Herbaceous perennial, with thin creeping rhizomes; stems mostly erect, often villous or with orange glands, 10–70 cm tall; basal leaves elliptic, lanceolate or ovate, 3–13 cm long, 1–3.2 cm wide; cauline leaves alternate, sessile, elongate-lanceolate to lanceolate; inflorescence a capitulum, 3–5 cm wide, single or in corymbiform groups; involucral bracts linear, 4–6 mm long, in 2 rows; ray flowers many (ca. 40–70), 1–1.5 cm long, yellow, twice as long as bracts; disc flowers 4–6 mm long, yellow; fruits linear-oblong achenes, ribbed, brown, with gray-white pappus (Fig. 2.25) (Zaurov et al. 2013; Malik et al. 2021).

Pollen features: Pollen unit: monad; pollen shape: spheroidal; pollen class: colporate; pollen aperture: sunken; aperture type: colporus; size (pollen unit): 10–25 µm (www.paldat.org).

Anatomical features: Toma et al. (2010) have reported an amphistomatic lamina; number of stomata per unit surface being higher in the upper epidermis; protective hairs completely resemble those found in the stem. The eco- anatomical observations reported by these workers underlines the presence of aerenchyma in the root/stem cortex. A plant has been collected from a salinized area, with a water table localized near the soil surface (Toma et al. 2010).

Part used: Underground parts, leaves

Mode of utilization: Infusion, decoction

Route of administration: Oral

Fig. 2.25 *Pentanema britannica* (L.) D.Gut. Larr., Santos-Vicente, Anderb., E.Rico & M.M.Mart.Ort.

Disease treated: Cystitis, diabetes, jaundice, respiratory catarrh, bone tuberculosis, rheumatism, hemorrhoids, vermifuge, hemostatic for uterine bleeding, to improve the appetite, an anti-inflammatory and astringent remedy (Zaurov et al. 2013; Malik et al. 2021).

Traditional uses: The plant is used for hiccups, nausea and excessive sputum together with other plants. The flowers are used to treat inflammation, bronchitis and intestinal diseases (Khan et al. 2010). Aerial parts or flower decoction are used for asthma in the traditional Chinese medicine as well as an expectorant (Khan et al. 2010). The flowers are used as stomachic, for hepatitis, as diuretic, laxative, antibacterial, carminative, tonic rapid-healer, and in tumors (Khan et al. 2010).

Recipes:

- Infusion or decoction of the underground parts used to treat bone tuberculosis, respiratory catarrh, diabetes, jaundice, cystitis, rheumatism and hemorrhoids as well as a vermifuge, to improve the appetite and hemostatic for uterine bleeding (Zaurov et al. 2013; Malik et al. 2021).
- Leaf infusion is taken as an astringent as well as anti-inflammatory agent (Zaurov et al. 2013).

Potential economic significance: None reported

2.21 *Pentatropis nivalis* (J.F.Gmel.) D.V.Field & J.R.I.Wood

Family: Apocynaceae
Synonym(s): *Asclepias nivalis* J.F.Gmel.; *Asclepias nivea* Forssk.; *Oxystelma caudatum* Buch.-Ham. ex Wall.; *Pentarrhinum fasciculatum* K.Schum.; *Pentatropis cynanchoides* R.Br. ex N.E.Br.; *Pentatropis fasciculata* (K.Schum.) N.E.Br.; *Pentatropis hoyoides* K.Schum.; *Pentatropis madagascariensis* Decne.; *Pentatropis nivalis* subsp. *madagascariensis* (Decne.) Liede & Meve; *Pentatropis rigida* Chiov.; *Pentatropis senegalensis* Decne.; *Strobopetalum carnosum* N.E.Br. (www.powo.science.kew.org).
English name: White milkweed
Flowering period: July–November
Distribution: Afghanistan, India, Iran, Oman, Pakistan, Palestine, Saudi Arabia, and Yemen (Malik et al. 2021).
Altitude: Up to 1300 m
Habit: Shrub
Ecological type: Xerophyte
Habitat: Scrambling over shrubs in dry bushlands, frequently on the seasonally flooded ground, or in riverine forests or thickets (www.powo.science.kew.org).
Phytochemicals: Alkaloids, cardiac glycosides, flavonoids, glycosides, phenols, resins, saponins, tannins, steroids, and terpenoids (Babre et al. 2020).
Potential biological activity: The plant extract has been evaluated for its antioxidant property (Babre et al. 2020).
Toxicity: Poisonous (Quattrocchi 2012).
Botanical description: Perennial twiner, nearly hairless, with tuberous roots; leaves very variable, ovate to linear, 2–4 cm long and 0.5–2 cm broad, more or less thick, hairless, with a short sharp point; flowers greenish-white, about 1.2 cm across, borne in 3–6 flowered cymes; flower divided almost to the base, with narrow elongate tepals; corona scales laterally compressed, vertically adnate to the backs of the anthers, with free tips and spurred bases, staminal tube short; seed-pods 5–8 cm long, about 1 cm broad, lance-shaped tapering into a beak, smooth, hairless; seeds with long hairs (Fig. 2.26) (www.powo.science.kew.org).

Fig. 2.26 *Pentatropis nivalis* (J.F.Gmel.) D.V.Field & J.R.I.Wood

Pollen features: None reported

Anatomical features: The seed coat pattern with a few mounds and papillae in between is characteristic; papillae covered by dense mounds and papillae in-between in the plant. The epidermal cells are randomly arranged, 4–6 gonal, and isodiametric in the seeds examined where they are elongated, slightly convex with a distinctly small protuberance in its central portion in the plant. The surface of the outer periclinal cell walls is tuberculated in the plant. These boundaries are well developed and indicated by channels in the seeds of the plant. These channels are smooth and distinctly deep shallow in the plant (Heneidak and Hassan 2005).

Part used: Whole plant

Mode of utilization: Raw, decoction, extract, juice, and paste

Route of administration: Oral, topical

Disease treated: Painful urination, acne, body pain, chronic fever, blood purifier, gonorrhea, dysuria, hepatoprotective, anti-smoking, heart tonic, astringent, mouth ulcer, muscle pain, leukoderma, throat sores, itching, kidney problems, backache, and menses problems.

Traditional uses: The plant is used for inflammation, leukoderma, biliousness (Ayurveda), gonorrhea, and emetic (Kirtikar and Basu 1987; Babre et al. 2020).

Recipes:

- Fruit is eaten raw for blood purification.
- Leaf decoction is used externally for itching and sore throat.
- Juice extracted from the leaves is used orally against jaundice.
- Fruit juice is taken orally against leukoderma and muscle pain.
- Root decoction is used as mouth wash for the treatment of mouth ulcer.
- Fruit paste with sugar is given to children twice a day as astringent.

Potential economic significance: The tubers of the plant are eaten (Ali 1983).

2.22 *Phoenix sylvestris* (L.) Roxb.

Family: Arecaceae	
Synonym(s): *Elate sylvestris* L.; *Elate versicolor* Salisb. (Malik et al. 2021).	
English name: Silver date palm, wild date palm, Indian date, sugar date palm (Malik et al. 2021).	
Flowering period: June–July (Malik et al. 2021).	
Distribution: Bangladesh, Bhutan, China, India, Myanmar, Nepal, and Pakistan (Malik et al. 2021).	
Altitude: Up to 1500 m (Malik et al. 2021).	
Habit: Tree	
Ecological type: Hydrohalophyte	
Habitat: From the plains to the coast in low-lying wastelands, scrub forests, and disturbed areas or the areas prone to periodic or seasonal inundation with water (Malik et al. 2021).	
Phytochemicals: Anthocyanins, phenolics, sterols, carotenoids, procyanidins, flavonoids, tannins, alkaloids, saponins, terpenoids, carbohydrates, amino acids, dietary fibers, essential vitamins and minerals (Sharma et al. 2016; Jain et al. 2018; Malik et al. 2021).	
Potential biological activity: Antioxidant, anthelmintic, antimicrobial, erythropoietic, antidiarrheal, analgesic, diuretic, anti-ulcer, antihypertensive and anti-diabetic activities (Jain et al. 2018).	
Toxicity: Acute toxicity and cytotoxicity (Mukherjee et al. 2014; Shajib et al. 2015; Jain et al. 2018; Malik et al. 2021).	
Botanical description: The stem of the plant is single or in cluster form, with some shoots up to 20 m high, to 52 cm wide, with permanent, diamond-shaped leaf base; leaves 4–6 m, sheathed with petiole up to 2 m, rachis 1–3 m; pinnae 200 per side of rachis, straight, not regularly arranged, spread in the diverse plane; mid pinnae up to 41 × 3 cm; male inflorescence erect, up to 2 m, with many rachillae, 31 cm; female inflorescences straight, become pendulum-like, up to 3 m; fruit flexible in shape, frequently oblong, 7.3 × 3.2 cm, brownish or blackish; endosperms homogeneous (Malik et al. 2021) (Fig. 2.27).	

Fig. 2.27 *Phoenix sylvestris* (L.) Roxb.

Pollen features: Pollen class monocolpate, shape: subdolate oroblate-spheroidal, ornamentation: tectum reticulate, polar length: 20.10–27.80 μm, equatorial diameter: 23.30–27.20 μm (Rashid and Perveen 2014).

Anatomical features: A single layer of rectangular or squarish epidermis, a thin layer of cuticles on their outer surface; 1–3 layers of hypodermis with squarish or oval cells, compactly arranged. In the hypodermal layers of its fruit some resinous substances have been reported. The conspicuous layer of tanniniferous cells is centrally placed, found in the multi-layered mesocarp, with a few scattered vascular bundles. Endocarp membraneous, 5–10 cells thick, papery and white (Biradar and Mahabale 1969).

Part used: Flower, leaves

Mode of utilization: Powder, decoction

Route of administration: Oral and topical

Disease treated: Hypertension

Traditional uses: The roots are evaluated for treating uterine tumors, swellings, menstruation and related problems, also used as tonic with butter and milk; used for backaches and sexual problems; as well as witchcraft. Domestic items made from the stems. Whole plants are broadly used for joint pains and fevers. In pneumonia and typhoid fevers and for hair dyeing the root extracts are used (Malik et al. 2021).

The plant is considered to cure fevers, heart complaints, loss of consciousness, abdominal complaints and constipation; plant sap too is a laxative, nutritious with cooling effect, central tender parts are used in the gonorrhea treatment; the root is also used in the treatment of helminthiasis, nervous debility and toothache (Jain et al. 2018).

Recipes:

- Dry flowers crushed and 2-tablespoons consumed with one cup of warm water daily in the early morning for hypertension for 14–15 days (Malik et al. 2021).
- A decoction made from 30 g of leaves by mixing these with 200 mL of water, used in a 100 mL cup 3 times a day for 14 days in hypertension patients (Malik et al. 2021).

Potential economic significance: According to Jain et al. (2018), *P. sylvestris* is a good source of natural antioxidants, nutritional supplements, the plant parts are widely used as functional foods in the food industry. It is highly popular among landscapers due to its low maintenance and beautiful appearance. As per Punjani (2002), local inhabitants in some areas of India use it in the construction of roofs of houses as supporting beams made from its trunk as well as for changing water path of water mill turbines. The forest dwellers use its stem and leaf at a large scale as construction materials (Kulkarni and Mulani 2004). The plant leaves are widely used for the preparation of hand fans and brooms as well as floor mats (Reddy et al. 2008; Kumar and Bhagat 2012; Rekha and Kumar 2014). Jain et al. (2018) also mention about the use of its long spines and other spines as toothbrush and traditional needles.

The flowers of wild date palm are evaluated worldwide in the preparation of sugar and alcohol. Its sweet fruits are consumed fresh as well as in the jelly form (Jain et al. 2018). The fat content of milk in the farm animals improves if they consume these on daily basis (Patil et al. 2014). In the Bengali cuisine, the plant sap is widely used for the preparation of sugar and traditional sweets (Ahmed 2007).

2.23 *Phragmites australis* (Cav.) Trin. ex Steud.

Family: Poaceae
Synonym(s): *Arundo australis* Cav.
English name: Common reed
Flowering period: July–September
Distribution: Afghanistan, China, Georgia, Iran, Iraq, Kazakhstan, Kyrgyzstan, Kuwait, Lebanon, Mongolia, Nepal, North Caucasus, Oman, Pakistan, Palestine, Russia, Saudi Arabia, Syria, Tadzhikistan, Transcaucasus, Turkiye, Turkmenistan, Uzbekistan, Vietnam, West Himalayas, West Siberia, Yemen (Malik et al. 2021).
Altitude: Up to 2400 m
Habit: Herb
Ecological type: Hydrohalophyte
Habitat: Lakes, rivers, ditches, swamps, fens, canal banks, and seashores (Davis 1985).
Phytochemicals: Phenolics, phytosterols, tannins, terpenoids, glycosides and flavonoids (Petropoulos et al. 2018; Samir et al. 2019; Malik et al. 2021).
Potential biological activity: Antioxidant, anti-melanogensis, and hepatoprotective (Petropoulos et al. 2018).
Toxicity: Heavy metal toxicity (Caparrós et al. 2022); and phytotoxicity (Rudrappa et al. 2009).
Botanical description: Perennial reed, with creeping rhizomes; culms erect, 1.5–3(6) m high; leaf-blades 20–60 cm (or more) long and 8–32 mm wide, glabrous, smooth beneath, the tips filiform and flexuous (sometimes stiff and pungent, see below); panicle 20–30(−50) cm long and 6–10(−15) cm wide, the lowest node usually few-branched, some of the branches bearing spikelets nearly to their base; spikelets 12–18 mm long, the rhachilla-hairs 6–10 mm long, copious, silky; lower glume 3–4.5 mm long; upper glume lanceolate, 5–9 mm long, sharply acute, usually apiculate; lowest lemma linear lanceolate to linear-oblong, 8–15 mm long; fertile lemmas very narrowly lanceolate, 9–13 mm long (Fig. 2.28) (www.efloras.org).

Fig. 2.28 *Phragmites australis* (Cav.) Trin. ex Steud.

Pollen features: Pollen type: apolar; pollen shape: prolate spheroidal; polar diameter: 27.5–32.5 μm; equatorial diameter: 27.5–30.0 μm; apertures: monoporate; sculpture type: psilate (El-Amier 2015).

Anatomical features: The transverse sections of mature plant leaf have been studies under light microscope. These possess a mesophyll of 12–15 cell layers; several layers of tightly appressed cells form the external part of the mesophyll, with limited intercellular spaces on both sides of the leaves. The mesophyll internal part shows irregularly shaped cells, larger than peripheral ones, separated by large intercellular spaces. Both leaf surfaces of leaves are studded with stomata, more numerous on the abaxial surface whereas, adaxial surface has large number of turgid bulliform cells extending from the leaf surface to the center of the mesophyll. A double bundle-sheath surrounds the vascular bundles (Antonielli et al. 2002).

Part used: Juice of the plant

Mode of utilization: Decoction, infusion

Route of administration: Oral

Disease treated: Diuretic, and antipyretic agent

Traditional uses: It has been used for centuries as a fodder plant in summer and the stems are traditionally harvested in winter as a raw material for crafts as well as for construction materials including roofing (Köbbing et al. 2013).

Recipes:

- A decoction of the juice of the plant is drunk as a cup daily for 3–4 weeks for diuretic effects (Arı et al. 2015).
- An infusion of the juice of the plant is used as an antipyretic agent (Arı et al. 2015).

Potential economic significance: In the wastewater management systems this plant plays a useful role. It is evaluated for the reduction of total suspended solids, biological oxygen demand and nitrogen from primary municipal waste waters in artificial wetlands (Gray and Biddlestone 1995). *Phragmites* reed stands found on the artificial as well as natural wetlands together with their associated microflora are excellent clean-up agents pollutant remediation together with sediment and other undesirable materials from water (www.cabi.org). This plant also helps in preventing soil erosion on the banks of rivers and channels (Bonham 1980).

For thatch and other traditional crafts this plant serves as a major harvested resource. In some countries it is also cut for pulp production, for example in the Danube Delta as well as Romania where we come across extensive reed swamp stands. The livestock also graze it, as such it is used as fodder, mainly when young. This information comes from the Volga delta, Russia and Kashmir (Langar and Bakshi 1990). In northern areas of China, the plant is grown as a crop (Li and Cao 1981).

According to Dely-Draskovits et al. (1992), the plant stands of this genus distributed around the wetlands play a major co-evolved wildlife support role, particularly in the temperate areas as vital parts of the wetland ecosystem and support wildfowl and other animals.

A report published by Arı et al. (2015) mentions that this plant is used for roofing, neck strips are used to transfer animals, as nests for migratory birds, fence construction, as animal shelters, for mat weaving used for wrapping and protecting corpses in graveyards, as well as in the tents for heat and sound insulation.

The species is used as animal food and has environmental uses for fuel and food (Ozturk et al. 2019; www.powo.science.kew.org).

2.24 *Plantago coronopus* L.

Family: Plantaginaceae

Synonym(s): *Asterogeum laciniatum* Gray; *Coronopus vulgaris* Fourr.; *Plantago lacustris* De Cresp.; *Plantago stellaris* Salisb. (www.powo.science.kew.org).

English name: Buckhorn plantain, Buck's-horn plantain

Flowering period: March–April

Distribution: Afghanistan, Iran, Iraq, Kuwait, Lebanon, North Caucasus, Pakistan, Saudi Arabia, Syria, Tadzhikistan, Turkiye, Turkmenistan, and Uzbekistan (Malik et al. 2021).

Altitude: Up to 800 m

Habit: Herb

Ecological type: Xerophyte

Habitat: Grassland, along rivers, in fresh and saline wetlands, and on coastal dunes (Weber 2003).

Phytochemicals: Flavonoids, phenolics, phenylpropanoid glycosides, and iridoids (Janković et al. 2012).

Potential biological activity: Antioxidant (Ceccanti et al. 2022).

Toxicity: Cytotoxicity (Galvez et al. 2003).

Botanical description: Perennial or annual, small, stemless herb, up to 15(−20) cm tall, hirsute; hairs more or less appressed; leaves oblong-lanceolate, 8–10(−12) cm long, 1(−1.8) cm broad, pinnatidentate to (bi-) pinnatifid; segments linear to lanceolate, acute; scapes 5–8 cm long, ascendent or erect, longer to sometimes shorter than the leaves; spikes narrow cylindrical; bracts ovate, acute to acuminate, narrowly margined, covered with short appressed hairs; anterior sepals narrow to broadly elliptic, 2–2.75 mm long, margin, ciliolate, carinate, posterior sepals ovate, broad, up to 3 mm long; corolla tube pilose; seeds 4–5 (Fig. 2.29) (www.efloras.org).

Pollen features: Pollen class: pantoporate; polar diameter: 24.20–30.44 μm; equatorial diameter: 19.69–24.94 μm; apertures: ectoaperture; ornamentation: verrucate (Klimko et al. 2004).

Fig. 2.29 *Plantago coronopus* L.

Anatomical features: Epidermal cells: rectangular with straight cell walls; stomata type: diacytic; trichomes: two non-glandular types: bottle-like and larger, long-stalked, multicellular trichomes; rarely, secretory trichomes could be observed (Andrzejewska-Golec 1992; Ianovici 2010; Cornara et al. 2022).
Part used: Leaves
Mode of utilization: Infusion, powder
Route of administration: Oral, topical
Disease treated: In the treatment of boils and wounds
Traditional uses: The plant is used for its analgesic, anti-inflammatory, antipyretic, anticancer and emollient properties, as well as to treat respiratory problems (Gonçalves and Romano 2016; Pereira et al. 2017; Cornara et al. 2022).
Recipes: • An infusion of the leaves of the plant is used for wounds (Ozturk et al. 2014). • A powder from the leaves of the plant is used for the treatment of boils (Ozturk et al. 2014).
Potential economic significance: The leaves of the plant are cooked as a vegetable in Balkan traditional cuisine. In France and Italy the leaves are mixed with other species to prepare salads, with a particular taste and crunchiness (Cornara et al. 2022).

2.25 *Plantago lanceolata* L.

Family: Plantaginaceae	
Synonym(s): *Arnoglossum lanceolatum* (L.) Gray; Lagopus lanceolatus (L.) Fourr.; *Plantago lanceolata* var. *lanceolata; Plantago sinuata* Lam. (Malik et al. 2021).	
English name: Narrow leaf plantain (Malik et al. 2021).	
Flowering period: March–August (Malik et al. 2021).	
Distribution: Afghanistan, China, East Himalayas, Iran, Iraq, Kazakhstan, Kyrgyzstan, Kuwait, Lebanon, Mongolia, Nepal, Pakistan, Palestine, Russia, Saudi Arabia, Syria, Tajikistan, Taiwan, Transcaucasus, Turkiye, Turkmenistan, Uzbekistan, West Himalayas, West Siberia, Yemen (Malik et al. 2021).	
Altitude: Sea level −3050 m (Malik et al. 2021).	
Habit: Herb	
Ecological type: Xerophyte	
Habitat: Seashores, sandy beaches, meadows, marshy ground, macchie, streamsides, coppices, roadsides, and waste places (Malik et al. 2021).	
Phytochemicals: Flavonoids, iridoid glycosides (especially catapol and its precursor aucubin), phenolics, terpenoids, tannins, alkaloids, some organic acids, mucilage, silica, zinc, and potassium salts (Marak et al. 2000; Shah et al. 2021; Malik et al. 2021).	
Potential biological activity: Used as an antibacterial agent for cuts, bruises and abscesses, topical anodyne, immunoregulatory, anti-inflammatory, antispasmodic, expectorant, emollient, astringent (Nichita et al. 2016; Malik et al. 2021). This plant has gained much popularity because of its antioxidant and antiviral activities in Turkiye and the Indian subcontinent (Shah et al. 2021). It is also reported to show anti-tumor and anti-ulcer activity, is anthelmintic, possesing hypotensive features (Rahamoz-Haghighi et al. 2021).	
Toxicity: Cytotoxic, and genotoxic (Çelik and Aslantürk 2006; Malik et al. 2021).	
Botanical description: A herb, perennial, short 31 (−61) cm long; leaves membrane type, narrowed lance-like to narrowed elliptical, 11–26 (−41) cm length, 2–4 (−6) cm wide, slight glabrous, piloted, attenuate, acute, base narrow into narrow petioles, nerve 6; scape erect, 16–31 (−61) cm length, sulcate, thinly cover with whitish hair; spike densely, conical cylinder, subglobose to globose, 1.2–3.1 (−5) cm long; bract broad oval, narrow caudate, 5–6 cm length, undulate, craniate; sepal 3.2–3.6 mm length, margin glabrous villose, sepal anterior connote, oval, bilobe, petal lobes narrowed oval; seeds 3, soft (Malik et al. 2021) (Fig. 2.30).	
Pollen features: Pollen unit: monad; colpus: tricolporate; exine sculpturing: reticulate; polar diameter: 16.7 (16.18–17.22) μm; exine thickness: 2.8 (2.7–2.9) μm (Khan et al. 2020).	
Anatomical features: The epidermal cells of the plant have an isodiametric shape. The plant has isolateral mesophyll with palisade parenchyma. The leaves are amphistomatic, and the stomata are anomocytic (Fig. 2.31). Glandular trichomes reported on both sides of the leaf epidermis. The plant has of an exclusive capitate glandular type is found trichome on both leaf surfaces, with epidermal basal cell, a stalk cell and apex comprised of multiple spindle glandular apical cells (de Souza Mesquita et al. 2017).	
Part used: Leaf, root, seed	
Mode of utilization: Paste, powder, decoction, extract, and raw	
Route of administiation: Oral, topical	
Disease treated: Rheumatism, fever, scorpion sting, hair problems, pulmonary diseases, scorpion sting, hair problems, peptic ulcer, malignant ulcers, parasitic worms, eye diseases, inflammations, hepatitis, diarrhea, cystitis, hay fever, and lung diseases.	

Fig. 2.30 *Plantago lanceolata* L.

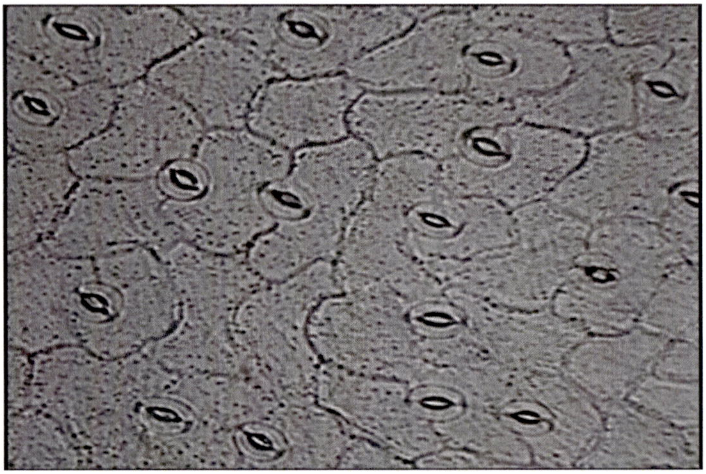

Fig. 2.31 Abaxial surface of *Plantago lanceolata* L.

Traditional uses: It is used to a large extent in the treatment of uterine cancer, pimples, homeostatic, tuberculosis, stomach disorders, asthma, bronchitis, diarrhea and blisters (Shah et al. 2021).

Recipes:

- Leaf paste is used externally to treat rheumatism.
- Root powder with *Marrubium vulgare* is taken orally for pulmonary diseases.
- Seed decoction is used to treat dysentery.
- Warm leaf is placed on skin to treat scorpion sting.
- Leaf extract is used to wash hairs for shining.
- Root extract is applied dermally to cure malignant ulcers.
- Seed extract is applied externally as eye lotion to remove circles.
- Roasted seeds are eaten orally for parasitic worms.

Potential economic significance: According to Ahatovic et al. (2020) and Pol et al. (2021), this species shows high tolerance towards eco-physiological conditions, therefore it may have good potential for evaluation in the recultivation of heavy metal contaminated soils. It has been reported to show good mutualistic relationships with arbuscular mycorrhizal fungi (Bennett et al. 2009). In view of the toxins present in its biomass of *P. lanceolata,* it is reported to contain special features of defense metabolites which protect it against herbivorous deterrent and pathogens, which can be oviposition stimulants for specialized insects (Fontana et al. 2009; Pol et al. 2021).

For the diet of pasture animals a mixture with another plant can serve as a richer source of nutrients due to a higher digestibility of the organic matter and metabolizable energy content (Somasiri et al. 2020; Pol et al. 2021).

The leaves of this species serve as a very good addition to livestock feed, as these are characterized by above-average health values; can also be used as a substitute for hay in sheep and a replacement for vegetables in animal diets (Pol et al. 2021).

These plants can be used to produce essential oil components for further evaluation in the food industry in food flavorings (Pol et al. 2021).

2.26 *Portulaca oleracea* L.

Family: Portulacaceae
Synonym(s): *Portulaca hortensis* Rupr.; *Portulaca officinarum* Crantz
English name: Common purslane, Purslain, Purslane, Wild purslane
Flowering period: May–September
Distribution: Afghanistan, Iran, Iraq, Kuwait, Lebanon, North Caucasus, Oman, Pakistan, Palestine, Saudi Arabia, Syria, Transcaucasus, Turkiye, Yemen (Malik et al. 2021).
Altitude: Up to 1500 m
Habit: Herb
Ecological type: Xerophyte
Habitat: In the open fields, on dunes, beaches, salt marshes, waste areas, eroded slopes, bluffs and riverbanks (www.cabi.org).
Phytochemicals: Alkaloids, flavonoids, phenolics, terpenes, monoterpene glycosides, coumarins, organic acids, fatty acids, amino acids, and vitamin E (Ozturk et al. 2021).
Potential biological activity: Antioxidant, neuroprotective, antimicrobial, antidiabetic, antioxidant, anti-inflammatory, antiulcerogenic, anticancer and antidepressant activities (Zhou et al. 2015; Ozturk et al. 2021).
Toxicity: Cytotoxicity (Chen et al. 2010).
Botanical description: Annual herbaceous and succulent plant, the height reaching up to 30 cm; the plant erect or decumbent; stems cylindrical, 2–3 mm in diameter, green or red; the stem glabrous except for leaf axils and scattered branches; leaves alternate or subopposite, usually flat and obovate, 1–5 cm long and 0.5–2 cm in diameter; wide or slightly notched at the apex, tapering at base, generally sessile and indistinctly petiolate; usually green in color and green with red margin; flowers small, orange-yellow, purple or white pink color, in clusters ranging from 2 to 5 groups or single at the ends of stems; the fruits approximately 4–8 mm long, as round shaped capsules, and each capsule contains a large number of seeds; seeds tiny, usually less than 1 mm in diameter, circular to egg shaped, with a flattened white connection point and vary in color from brown to black (Fig. 2.32) (Ozturk et al. 2021).
Pollen features: Pollen shape: spheroidal; polar length: 48.2–58.5 μm, equatorial length: 48.2–58.5 μm; exine ornamentation: microechinate-anulopunctate; aperture type: pantocolpate (Shamso and Toshiyuki 2012).
Anatomical features: The epidermal cells are irregular in shape and cell wall outlines are strictly wavy on both leaf surfaces. Leaves are amphistomatic, and on both surfaces, stomata are paracytic type. Trichome is absent on both the epidermal surfaces. Star-shaped crystals of calcium oxalate are present on both the leaf surfaces (Pal and Rahaman 2014).
Part used: Whole plant, leaves
Mode of utilization: Juice, cooked, paste, maceration
Route of administration: Oral, topical
Disease treated: Intestinal worm infestation, bilious dysentery, burn wounds, inflamed areas, boils on the scalp and headaches, dysuria, high fever, gastritis, polymenorrhagia, bleeding piles, and hemoptysis (Sultana and Rahman 2013).

Fig. 2.32 *Portulaca oleracea* L.

Traditional uses: It has been used in traditional Chinese medicine orally for the treatment of dysentery with bloody stools, and externally for swellings, erysipelas, eczema, snake- and insect bites, abnormal uterine bleeding and hemorrhoid bleeding, as well as sores (Chen et al. 2009; Iranshahy et al. 2017). The leaves and seeds are used in the blood purification as well as curing cardiovascular complaints and circulatory diseases and dental problems in Nepal. A paste is applied to the teeth and gums to cure toothache after preparing it from the fruits and seeds (Joshi and Joshi 2000). The findings reported by Nadkarni (1996), Belcheff (2012) and Iranshahy et al. (2017) reveal that the aerial parts are recommended as anti-scorbutic, refrigerant, mild diuretic, wound healer and anti-rheumatic in Philippines; and in the Ayurvedic medicine it is used to cure scurvy, asthma, leprosy, hemorrhoids, gastric inflammation, diseases of lungs, liver, kidneys, bladder and bowels and spitting of the blood. In the treatment of burning sensation the plant juice from this species is recommended. In the treatment of diseases of kidney, bladder and lung both plant and the seeds are used; externally used to alleviate burns, scalds and skin diseases (Belcheff 2012; Iranshahy et al. 2017).

The fresh aerial parts of the plant are considered valuable in the treatment of urinary and digestive problems in Pakistan since years. In the treatment of bladder ailments such as dysuria, its juice is reported to be useful due to its diuretic properties. It is also used to treat gastrointestinal problems such as diarrhea and dysentery. As per Ullah et al. (2013), its seeds are demulcent, diuretic and vermifuge. The fresh leaves are slightly heated and applied topically on swelling joints and the stem extract is applied on skin to cure burning sensation. This species is believed to be depurative, febrifuge, cardiac stimulant and good for treatment of sores, burns, earache, skin infections and coughs (Abbasi et al. 2015). In Afghanistan, The seeds are used as an antidiarrheal and in the treatment of throat infection in Afghanistan (Younos et al. 1987). As per El-Ghonemy (1993), the aerial parts are considered as a useful febrifuge in United Arab Emirates and Oman. The seeds are reported to be evaluated as a blood purifier and an aphrodisiac in Jordan (Lev and Amar 2002; Iranshahy et al. 2017). The leaves are used to cure urinary disorder, diabetes, headache, diarrhea, ulcers and wounds in Turkiye. The plant is consumed in Saudi Arabia for the treatment of liver and gastrointestinal problems, and inflammatory diseases (Al-Asmari et al. 2014; Iranshahy et al. 2017).

Recipes:
- Juice of the plant (30 g) with sugar is administered orally and is beneficial in intestinal worm infestation.
- A whole plant cooked with butter and onion is beneficial for bilious dysentery.
- Paste of the whole plant made with rose oil relieves headache.
- This paste is also used on burn wounds, inflamed areas, orchitis and pleurisy.
- Juice with rose oil is effective to cure boils on the scalp and headaches.
- Juice of leaves and branches (60 g) is commonly useful in dysuria.
- Orally, macerated leaves are given in high fever.
- The whole plant or its juice is useful in gastritis. It is hemostatic thus used in polymenorrhagia, bleeding piles, hemoptysis (Sultana and Rahman 2013).

Potential economic significance: The species is in fact known to the humans for the last 2000 years in different parts of the globe. It has been traditionally consumed as food in several regions of the world, and in the tropical Asian and Mediterranean countries it is added to soups and salads (Altay and Çelik 2011; Altay and Karahan 2012; Altay et al. 2015; Ozturk et al. 2021), and at the same time serves as a source of gum with emulsification characteristics for its further evaluation in the food industry (Garti et al. 1999; www.cabi.org).
It is highly tolerant to chloride- as well as sulphate-dominated saline conditions; a moderate selenium accumulator but a valuable vegetable crop for human consumption and for livestock forage (Bianco et al. 1998; Ozturk et al. 2019).

2.27 *Prosopis juliflora* (Sw.) DC.

Family: Fabaceae

Synonym(s): *Acacia juliflora* (Sw.) Willd.; *Algarobia juliflora* (Sw.) Heynh.; *Entada juliflora* (Sw.) Roberty; *Mimosa juliflora* Sw.; *Neltuma juliflora* (Sw.) Raf. (www.powo.science.kew.org).

English name: Mesquite

Flowering period: March–June

Distribution: Bahrain, Bangladesh, Cambodia, China, India, Indonesia, Iran, Iraq, Kuwait, Myanmar, Nepal, Oman, Pakistan, Philippines, Qatar, Saudi Arabia, Sri Lanka, Taiwan, Thailand, United Arab Emirates, Vietnam, West Himalayas, and Yemen (EMPPO 2019).

Altitude: Up to 500 m

Habit: Tree

Ecological type: Xerohalophyte

Habitat: Coastal areas, forest, grasslands, wetlands, and ruderal areas (EMPPO 2019).

Phytochemicals: Alkaloids, flavonoids, phenolics, tannins, terpenoids, saponins, coumarin, flavonol glycoside, glycoside and aminoacids (de Brito Damasceno et al. 2017).

Potential biological activity: Antimicrobial, antioxidant, antimalarial, antitumor, anthelmintic, anti-emetic and cholinesterase inhibiting activities (de Brito Damasceno et al. 2017).

Toxicity: According to the data published by de Brito Damasceno et al. (2017), its toxic effects are well reported in animals who have free access to the plant as a food source and this leads to a muscle dysfunction. The juliprosopine and juliprosine alkaloids are possibly mainly responsible for the neurotoxic damage in animals following consumption of this plant as per Silva et al. (2013); a possible mechanism of action is supposed to be based on induction of mitochondrial dysfunction, gliosis and autophagic vacuolation. The leaf suspensions made from this plant during summer and winter seasons are reported to show significant larval toxicity (de Brito Damasceno et al. 2017).The plant has also been reported to have a cytotoxic effect (Hughes et al. 2006; de Brito Damasceno et al. 2017).

Botanical description: Tree 3–12 m tall; wood hard, branches cylindrical, green, more or less round- or flat-topped, somewhat spiny with persistent, green (sometimes glaucous or greyish, not reddish) foliage, glabrous or somewhat pubescent or ciliate on the leaflets; spines axillary, uninodal, divergent, paired, or solitary and paired on the same branches, sometimes absent, not on all branchlets, measuring 0.5–5.0 cm long, being largest on strong, basal shoots; leaves bipinnate, glabrous or pubescent, 1–3 pairs of pinnae, rarely 4 pairs; petiole plus rachis (when present) 0.5–7.5 cm long; pinnae 3–11 cm long; leaflets 6–29, generally 11–15 pairs per pinna, elliptic-oblong, glabrous or ciliate, rarely pubescent, approximate on the rachis or distant a little more than their own width, herbaceous to submembranous, emarginated or obtuse, pinnate-reticulately curved; leaflets 6–23 mm long 1.6–5.5 mm wide; racemes cylindric, 7–15 cm long, rachis puberulent; florets as usual, greenish-white, turning light yellow; legume straight with incurved apex, sometimes falcate, straw-yellow to brown, compressed, linear with parallel margins, stalked and acuminate, 8–29 cm long 9–17 mm broad 4–8 mm thick; stipe to 2 cm; endocarp segments up to 25, rectangular to subquadrate, mostly broader than long; seeds oval, brown, transverse (Fig. 2.33) (EMPPO 2019).

Pollen features: Pollen class: tricolporate, shape: sub-oblate, polar: 25.13–28.7 μm, equatorial length: 28.7–34.1 μm (Perveen and Qaiser 1998).

Anatomical features: It has a bipinnately compound leaf with entire margin, apex is blunt, base round, venation reticulate, epidermal cells being thick and straight walled, paracytic type stomata, and large mucilage cavities in the mesophyll tissue (Fig. 2.34) (Robertson et al. 2010).

Fig. 2.33 *Prosopis juliflora* (Sw.) DC.

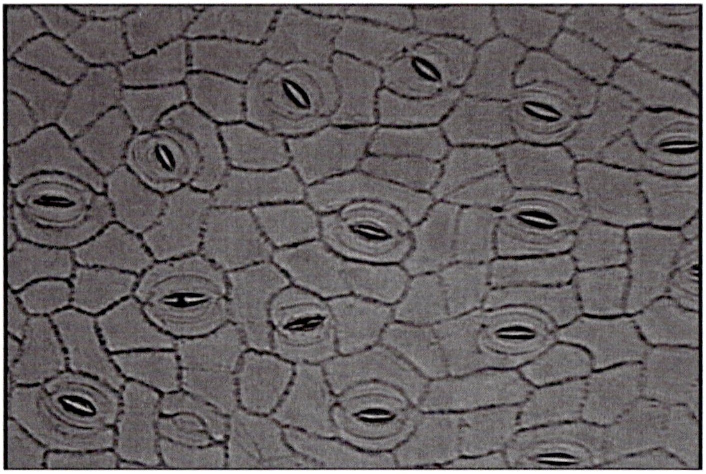

Fig. 2.34 Abaxial surface of *Prosopis juliflora* (Sw.) DC.

Part used: Bark, fruit, leaf, flower, stem, gum

Mode of utilization: Decoction, powder, extract, paste, raw

Route of administration: Oral, topical

Disease treated: Dysentery, pneumonia, influenza, improves memory, bladder tonic, paralysis, testis swellings, skin disorders, bones weakness, gall stones, body swellings, increase lactation, conjunctivitis, hay fever, hernias, and acnes.

Traditional uses: Its use has been reported in different studies. The drugs prepared from different parts of the plant by infusion, decoction and syrup methods are used for treatment, especially against asthma, bronchiti, conjunctivitis, digestive disturbances, skin lesions, and relief of painful teeth (Hebbar et al. 2004; Tene et al. 2007; Agra et al. 2008; Singh and Swapnil 2011; Kayani et al. 2014).

Recipes:

- Stem bark is boiled in water for 15 minutes which is used twice in a day to treat asthma.
- Tea prepared using dried pods is used for gastric problems.
- Fruit decoction is used to treat hay fever.
- Leaf powder with honey is applied externally to eyes to cure conjunctivitis.
- Flower extract is given to mother for increase lactation.
- Boiled stem paste is applied dermally on effected areas to reduce body swellings.
- Flower decoction is taken orally to cure hernias.
- Gum is applied dermally to open wounds.

Potential economic significance: EMPPO (2019) reported that the plant has been principally used as wood for poles, posts, fuel and sawn timber. The pods are consumed as fodder as well as human food sources. Several other tree products reported such as; wood as biofuel for electricity generation, exudate gums, fibers, tannins, leaf compost honey from the flowers, medicines from various plant parts, and chemical extracts from the wood or pods; also widely planted for soil conservation, in hedgrooves, as an urban and general amenity tree (EMPPO 2019).

2.28 *Saccharum spontaneum* L.

Family: Poaceae	
Synonym(s): *Imperata spontanea* (L.) P.Beauv.; *Saccharifera spontanea* (L.) Stokes (Malik et al. 2021).	
English name: Wild sugarcane, Kans grass (Malik et al. 2021).	
Flowering period: July–September (Malik et al. 2021).	
Distribution: Afghanistan, Bangladesh, Cambodia, China, East Himalayas, India, Kazakhstan, Kyrgyzstan, Lebanon, Myanmar, Nepal, Oman, Pakistan, Palestine, Philippines, Saudi Arabia, Sri Lanka, Syria, Taiwan, Thailand, Tibet, Turkmenistan, Uzbekistan, Vietnam, West Himalayas, and Yemen (Malik et al. 2021).	
Altitude: Up to 1800 m	
Habit: Herb	
Ecological type: Hydrohalophyte	
Habitat: Common along riverbanks, roadsides and railroads, on waste ground, and along the banks of lakes and ponds (Holm et al. 1997; Malik et al. 2021).	
Phytochemicals: Flavonoids, phenolic, tannins, and polyphenols (Sathya and Kokilavani 2013; Malik et al. 2021).	
Potential biological activity: The plant is recommended in the treatment of respiratory diseases as well as gynecological troubles. The roots in the Ayurveda system are used as galactagogue as well as diuretic, also evaluated as aphrodisiac, purgative, tonic, diuretic, emollient, as astringent, refrigerant, in the treatment of dyspepsia, burning sensation, and piles (Khalid et al. 2011).	
Toxicity: Low toxicity (Chaudhary et al. 2012; Malik et al. 2021).	
Botanical description: It is a tall, perennial rhizomatous plant; culm 3–5 m in height; leaf blades 0.6–3 m long, up to 7.6 mm broad, slowly tapering towards the base, glaucous, narrowly winged, present on all sides of the petiole; panicles 26–41(−61) cm in length, peduncles typically hairy; racemes 4–16 cm long, longer than the branches that support it, internodes and pedicel hirsute; spikelets also hirsute, 2.6–6(−8) mm in length, silky whitish hairs present in the callus; 3–4 times as longer as spikelet; glumes equal, in lower positions sub coriaceous, glabrous below, margin are ciliate on upper side; lemmas of lower sides are lance-like, margins ciliate; upper lemmas narrow, with shorter awns (Malik et al. 2021) (Fig. 2.35).	
Pollen features: Pollen type: 1-porate annulate; pollen size: 36 (31–40) μm; exine thickness: 1.5–2 μm (Chaturvedi and Datta 2001).	
Anatomical features: Distinct bulliform cells found in the leaves, which allow the leaf to fold or roll (Fig. 2.36). The leaf vascular bundles are closed, numerous, collateral and diffusedly distributed, larger in size and surrounded by a sheath of fibers. In the stem cross-section these are more or less circular, ground tissue being parenchymatous, undifferentiated into cortex and pith; bundles are collateral, usually surrounded by a sclerenchymatous sheath (Abd El-Gawad and El-Amier 2017).	
Part used: Leaf, root, stem, flower	
Mode of utilization: Juice, extract, decoction, paste	
Route of administration: Oral, topical	
Disease treated: Tuberculosis, kidney stones, skin boils, wound healing, toothache, hysteria, leukorrhea, scabies, mouth gums, throat pain, influenza, abscesses, and earache.	
Traditional uses: According to Kumar et al. (2010) and Pandey et al. (2015), the roots are used as an aphrodisiac, tonic, purgative, lithotriptic, diuretic, astringent, emollient, refrigerant and in the treatment of sexual weakness, piles, respiratory disorders, burning sensation, dyspepsia, and gynecological troubles. A fresh stem juice is good for the treatment of mental illnesses and is being evaluated by different tribes in India (Pandey et al. 2015).	

Fig. 2.35 *Saccharum spontaneum* L.

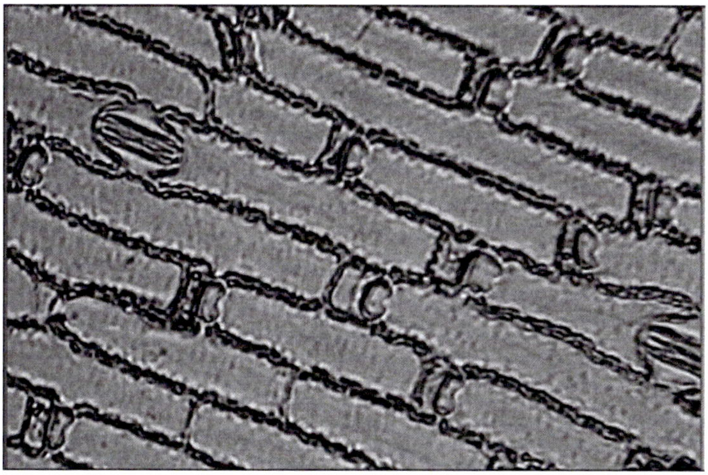

Fig. 2.36 Abaxial surface of *Saccharum spontaneum* L.

Recipes:

- Leaf juice with milk is taken orally for treatment of cough in tuberculosis.
- Root extract is taken as cooling agent during the treatment of jaundice.
- Leaf decoction is taken orally to break kidney stones.
- Stem paste is applied dermally as a poultice on skin boils and wounds.
- Leaf paste is applied on teeth to get relief from toothache.
- Extract of inflorescence is applied externally to cure mouth gums.

Potential economic significance: For a production of different grades of paper, especially grease-proof paper culm of this plant is a good source of pulp. The leaves are well evaluated as thatching material and to support their livelihood, local inhabitants in many areas use it for making brooms, baskets, mats, ropes, in huts, etc. (Pandey et al. 2015). According to Pandey et al. (2015) and Ozturk et al. (2019), it is good as a fodder for goats and camels in the juvenile stage, and is suitable for the production of silage. Wapakala (1966) has reported long back that its slow decomposition rate makes it an excellent mulching material.
The plant is rich in carbohydrates in Its cell walls (67.85% on a dry solid basis), making it a novel and suitable substrate for ethanol production (Scordia et al. 2010; Chandel et al. 2011). A fast-growing plant with good biomass and flowers are rich in fibers, which are white/purplish silky and have better strength and fineness, distinctly different in appearance from other types of fibers such as cotton, jute, flax, ramie and hemp (Pandey et al. 2015; Malik et al. 2021).
The ecological importance of this plant species is very high as it is effective against soil-erosion, due mainly to its extensive rhizome network (Pandey et al. 2015; Malik et al. 2021). Moreover, it also has religious importance in India (Pandey et al. 2015).

2.29 *Salsola imbricata* Forssk.

Family: Amaranthaceae

Synonym(s): *Caroxylon foetidum* Moq.; *Caroxylon imbricatum* (Forssk.) Akhani & Roalson; *Chenopodium baryosmum* Schult.; *Salsola baryosma* (Schult.) Dandy; *Salsola foetida* Delile ex Spreng.; *Salsola foetida* var. *glabrescens* Maire; *Salsola foetida* var. *scopiformis* (Maire) Maire; *Salsola imbricata* var. *imbricata* (www.efloras.org).

English name: Saltwort

Flowering period: August–October

Distribution: Afghanistan, India, Iran, Pakistan, the Arabian Peninsula (www.efloras.org).

Altitude: Up to 1500 m

Habit: Shrub

Ecological type: Xerohalophyte

Habitat: In naturally disturbed habitats such as eroded slopes, dry riverbeds and sea-shore cliffs (www.efloras.org). In addition, it is growing deserts and arid regions.

Phytochemicals: Alkaloids, flavonoids, phytosterols, saponins, tannins and phenolics (Andhiwal and Kishore 1984; Khan et al. 2003; Janbaz et al. 2021).

Potential biological activity: The crude extract of the plant is known for its tyrosine inhibitory potential, antioxidant and spasmolytic action, vermicidal, antibacterial, male contraceptive, diuretic, antispasmodic and broncho-relaxant, analgesic, anti-inflammatory, and antipyretic effects (Khan et al. 2003; Javed and Jabeen 2021).

Toxicity: An animal cannot consume much of this plant because of its high salt content, therefore restricting its consumption, an overconsumption can lead to mineral toxicity in animals (Altay and Ozturk 2021).

Botanical description: Subshrub or small shrub, 0.3–1.2(2) m high; stems and lower leaves densely covered with spreading or ascending, curved, warty, short and to 1.5 mm long straight hairs, in young stage reddish-brown, later glabrescent, bracts and bracteoles only sparsely and shorter hairy, glabrescent; stem to 2 cm thick, nodulated, grey or reddish, densely branched in upper part; branches, ascending, the upper often spreading, in terminal parts milky white or grey, with regularly arranged, spreading lateral catkin-like spikes; spikes 5–35 cm × 2–3(4) mm, condensed or interrupted; leaves succulent, with flat to a convex surface, linear to narrowly triangular in lower parts, in upper scale-like, cucullate, broadly ovate to circular in outline, obtuse, (1.5)2–7 × 0.5–1.3 mm; leaf axils of long shoots with small rosulate scaly leaves; bracts and bracteoles subequal in shape and size, similar to upper leaves, 0.7–1.5 × 0.9–1.5 mm, usually glabrous, shorter than perianth; tepals ovate to ligulate, (1)1,2–1.5(2) mm long, the outer 0.8–1.2 mm wide, apex obtuse, usually crenulate, 1-veined, transverse line at 1/2, green blotch small, obtusely triangular, margins of outer tepals ciliate up to 1/2 with almost smooth curved hairs, blotch area sometimes with scattered ascending warty hairs; anthers 0.7–1.1 mm long, including the 0.1–0.15 mm long obtuse appendage, divided for 1/2–2/3; filaments 1.5–2 × mm long, at base 0.2 mm wide; disc absent; Style cylindric, 0.4–0.6 × 0.2 mm; stigmas 2, 0.4–0.7 mm long, flat up to the apex, revolute, inside long papillose; fruiting perianth 3–5 mm, the wings silky, translucent, subequal or the 2 inner narrower; upper part of tepals suberect or incurved, forming an obtuse cone or a semi-globular dome; seed semi-globular, flattened at top, 1–1.2 mm (Fig. 2.37) (www.efloras.org).

Pollen features: Pollen type: apolar; pollen shape: spheroidal; pollen ornamentation: scabrate; polar diameter: 16.0–18.5 μm; exine thickness: 1.25–2.25 μm (Nazish et al. 2019).

Fig. 2.37 *Salsola imbricata* Forssk.

Anatomical features: Nazish et al. (2020) report that amphistomatic leaves are observed in the plant, with paracytic type of stomata. The types of trichome have been reported glandular, non-glandular and conical (Nazish et al. 2020).

Part used: Leaf, flower, stem, seed

Mode of utilization: Ash, paste, extract, decoction, juice

Route of administration: Oral, topical

Disease treated: Abdominal distention, constipation, dyspepsia, ophthalmia, skin diseases, blood clotting, abortion, bone pain, asthma, git worms, labor weakness, chest diseases, pulmonary congestion, brain tonic, and kidney stones.

Traditional uses: The plant is used in gastrointestinal disorders including poor digestion, vomiting, piles, dyspepsia, and abdominal distention. It is also used in the treatment of headaches, migraine, vertigo, scabies, eruption, and wounds (Khan et al. 2003; Ahmad et al. 2014; Wariss et al. 2017; Javed and Jabeen 2021).

Recipes:

- Aerial parts ash with sugar is taken orally for abdominal distension.
- Leaf paste is used as poultice to treat ophthalmia.
- Flower extract is used dermally for skin fairness.
- Stem decoction is used orally to cure kidney stones.
- Leaf juice is applied externally on wounds for blood clotting.
- Flower paste is applied externally to treat bone pain.
- Seed decoction is used orally for abortion.

Potential economic significance: The plant is widely distributed in Punjab, northern Gujrat, Rajasthan, and the Cholistan, where it is used to prepare "sajji," a crude form of carbonate of soda (Khan et al. 2003).

A promising camel fodder in the hypersaline, semiarid, and arid areas across the globe (Hanif et al. 2017; Ozturk et al. 2019; Altay and Ozturk 2021). The plant is used as a potential candidate for the rehabilitation of rangelands (Hanif et al. 2017).

2.30 *Salvadora oleoides* Decne

Family: Salvadoraceae	
Synonym(s): *Salvadora stocksii* Wight	
English name: Toothbrush tree	
Flowering period: March–April	
Distribution: India, Iran, Pakistan, Yemen	
Altitude: Up to 1000 m	
Habit: Tree	
Ecological type: Xerohalophyte	
Habitat: Roadsides, canal and riverbanks, dryland salinities and saltmarshes, dry mountains, and deserts and semi-deserts (Iqbal et al. 2021).	
Phytochemicals: Alkaloids, steroids, glycosides, saponins, tannins, triterpenes, mucilage, carbohydrates, fatty acids, and essential oils (Arora et al. 2014).	
Potential biological activity: Anti-inflammatory, analgesic, anti-ulcer, anthelmintic, antimicrobial, and antidiuretic activities (Arora et al. 2014).	
Toxicity: The seed oil of the plant shows 100% toxicity to *Anopheles stephensi* at 0.01% (www.worldagroforestry.org).	
Botanical description: The plant is a small tree, attaining 6–9 m height; trunk short, often twisted or bent, up to 2 m in diameter; branches drooping, numerous, stiff, often swollen at forks; bark grey or whitish-grey; leaves glaucous, linear-or ovate-lanceolate, coriaceous and somewhat fleshy, dark greenish-yellow when young, grey when mature; flowers sessile, greenish-white, minute in paniculate spikes, often clustered; calyx cup-shaped, in 4 rounded, obtuse lobes; Fruit a drupe, globose, about 6 cm in diameter, usually yellow when ripe, dark brown or red when dry; seeds greenish-yellow, about 3 mm in diameter (Fig. 2.38) (www.worldagroforestry.org).	
Pollen features: Pollen class: 3-zonocolporate; shape: sub-prolate to prolate - spheroidal; polar diameter: 8.41–15.41 μm; equatorial diameter: 8.41–12.6 μm (Perveen and Qaiser 1996).	
Anatomical features: Much variation is seen in various tissue systems; collenchyma, storage parenchyma, shape, size of internal oil glands, the epidermis and epidermal appendages and arrangement of vascular tissue, stomatal size, sclerification intensity, shape and orientation. Additional parenchymatous layers found in the desert populations under multilayered epidermis, phloem fully encircles the xylem by broad metaxylem vessels, in and outside vascular tissue intensive sclerification is visible, with numerous and small sized stomata on abaxial leaf surface. The salt-affected areas have populations which show multilayered epidermis, large storage parenchyma, enlarged palisade parenchymatous cells, greater stem cross-sectional area and sclerified vascular bundles. Sclerified vascular bundles and dense hairiness seen on the leaf surfaces of roadside populations (Iqbal et al. 2021).	
Part used: Bark, stem, flower, leaves	
Mode of utilization: Decoction, juice, powder, ash	
Route of administration: Oral, topical	
Disease treated: Purgative, skin diseases, blister, stomach ulcer, spleen enlargement, female sterility, sciatica, pain killer, hepatitis, gum diseases, eye infection, liver diseases, abdominal worms, blood cancer, and flu.	

Fig. 2.38 *Salvadora oleoides* Decne

Traditional uses: According to Qureshi et al. (2010), Naresh et al. (2013) and Saleem et al. (2020), the leaves are used for the treatment of enlarged spleen and fever as well as for relieving cough. The stem bark is alexipharmic and a stimulant (Verma 2016). It is useful in the treatment of hypoglycemia, rheumatic pain, tumors, bronchitis, and piles (Galati et al. 1999; Naresh et al. 2013; Garg et al. 2014; Saleem et al. 2020).

Recipes:

- Stem bark decoction is purgative.
- Flower juice is used to treat skin ulcer and blisters.
- Decoction of young dried stem branches is used orally to get relief from body pain.
- Powder of dry flowers is taken orally with water for stomach ulcer and diuretic.
- Flower decoction is taken orally for spleen enlargement.
- Stem ash is used externally to treat sciatica.
- Leaf decoction is taken orally to cure female sterility.

Potential economic significance: In the arid zones, it is extensively used as fuel and fodder; being adaptable to a variety of soil and geographical conditions such as; saline/sodic or alkaline soils, hard rocky foothills, pure or loamy sand (Nafees et al. 2019; Iqbal et al. 2021). Ecologically it acts as a windbreak in desert areas and stabilization of sand dunes (Khan 1994). Leaves and young branches of the trees are favorite fodder for camel and other livestock (especially goats and sheep) in the region due to their highwater content (15–36%). The tree is an important source of fuelwood for the local inhabitants. The light-yellow wood is used for building purposes, for agricultural implements, making Persian wheels, and boats (Arora et al. 2014; Barman et al. 2018). Fruits are sweet, consumed by the local people, and are a rich source of calcium amounting to about 15 times that present in wheat (Rathore 2009; Barman et al. 2018). Fruits are also mixed in cattle feed to enhance lactation (Khan 1994). Seeds contain non-edible oil and large lauric and myristic acids and are thus used for making soap and candles (Barman et al. 2018).

2.31 *Solanum surattense* Burm. f.

Family: Solanaceae

Synonym(s): *Solanum jacquini* Willd. (Malik et al. 2021).

English name: Yellow Berried Nightshade (Malik et al. 2021).

Flowering period: Mostly throughout the year (Malik et al. 2021).

Distribution: Afghanistan, Bangladesh, China, East Himalayas, India, Iran, Myanmar, Nepal, Oman, Pakistan, Saudi Arabia, Sri Lanka, Taiwan, Thailand, Vietnam, West Himalayas, and Yemen (Malik et al. 2021).

Altitude: Up to 1300 m (Malik et al. 2021).

Habit: Herb

Ecological type: Xerophyte

Habitat: Along the roadsides and dry wastelands (Malik et al. 2021).

Phytochemicals: Alkaloids, flavonoids, phenols, saponins, steroids and triterpenoids (Malik et al. 2021).

Potential biological activity: antimicrobial, anihelmenthic, antihyperglycemic, hypolipidemic, cardiovascular protective effect, antiulcer, wound healing, uriolithiatic, and antifertility (Tekuri et al. 2019; Malik et al. 2021).

Toxicity: Metal toxic (Deng et al. 2010; Malik et al. 2021).

Botanical description: A prickly, prostrate, and diffused herb; prickles 16 mm in length, yellowish; stems and branches stellate, glabrous when mature; leaves 31–81 × 26–51 mm wide, elliptical-long, deep lobed to sinuate, above greenish; unequal lobes, acute or obtuse, lobulated or toothed; flowers 3–5, purplish; peduncles cymose; peduncles 11–21 mm in length; calyx lobe 6 mm long, prickly, acute; petals 3–2.9 cm wide; lobes 11–13 mm in length, triangular, ovate; anther 7.6 mm long, elongate; berries globose, 16–21 mm broad; seed discoidal, faintly reticulate (Malik et al. 2021) (Fig. 2.39).

Fig. 2.39 *Solanum surattense* Burm. f.

Pollen features: Pollen type: tricolporate, pollen shape: subspheroidal, polar axis: 16.25–17.50 μm, equatorial length: 27.50–29.25 μm (Ashfaq et al. 2020).

Anatomical features: According to Bibi et al. (2015), the first row of cells in the leaves is epidermis, followed by a wide cortex and endodermis; radial vascular bundles are clear, abundant water retention and conservation tissues; there is a parenchymatous cortex in the stem storing large amount of water, with radially arranged vascular bundles; pith is large and parenchymatous; in the roots epidermal layer has closely arranged elongated cells, the thick parenchymatous cortex stores enough water; the vascular bundles are radial, protoxylem towards the center and metaxylem towards the cortex.

Part used: Root, whole plant (Malik et al. 2021).

Mode of utilization: Powder, decoction (Malik et al. 2021).

Route of administration: Oral

Disease treated: Throat allergies, and curing cough

Traditional uses: As per Sharma et al. (2010), the roots have been used as a tonic for lactating mothers. The data published by Parmar et al. (2010) mentions that in the Dhenkanal district of Orissa (India) Kondh tribes are using the mature fruit hot aqueous extract as a traditional medicine for the treatment of diabetes mellitus. The locals in Manipur (India) are using fruits as a folk medicine to treat inflammatory problems as well as throat infection. According to Madhavi et al. (2014) and Malik et al. (2021), entire plant is used for treating leprosy, dropsy, and cough. The root paste is being evaluated by the Mukunda tribes (Rajasthan, India) for a treatment of hernia (Pandey et al. 2018). For the treatment of piles root poultice is used as a traditional medicine in many villages of South India (Tekuri et al. 2019). Panday (2004) has reported that the fumigated seeds with mustard oil are used as an excellent remedy for the treatment of tooth pain, dental caries and pus formation, together with the associated swelling of gums. According to the data published by Rahman et al. (2003), Pingale (2013), and Malik et al. (2021), the stem, flowers, and fruits are used as relief in the burning sensation in the feet and seeds are used as a remedy for cough and asthma.

Recipes:

- For treating throat allergie, sundry root is powdered (60 g), one tablespoon used daily with 2 glasses (200 mL) of water for 8–9 days, 3 times a day (Malik et al. 2021).
- A decoction is made from the whole plant (50 g) boiled with 200 mL of water for 6–7 min; 2 glasses (200 mL) of this are taken for 6–8 days for curing cough (Malik et al. 2021).

Potential economic significance: Not reported

2.32 *Suaeda fruticosa* Forssk. ex J.F.Gmel.

Family: Amaranthaceae

Synonym(s): *Dondia fruticosa* (Forssk. ex J.F.Gmel.) Druce; *Lerchia fruticosa* (Forssk. ex J.F.Gmel.) Medik.; *Salsola frutescens* Forsyth f.; *Schoberia fruticosa* (Forssk. ex J.F.Gmel.) C.A.Mey. (www.powo.science.kew.org).

English name: Shrubby Seablite

Flowering period: March–May

Distribution: Afghanistan, India, Iraq, Iran, Jordan, Pakistan, Palestine, Saudi Arabia and Yemen

Altitude: Up to 2000 m

Habit: Shrub

Ecological type: Xerohalophyte

Habitat: In flooded alluvial areas, dried areas, salt marshes, salt flats, and coastal areas (Saleem et al. 2021).

Phytochemicals: Chromo alkaloids (especially betacyanins and betaxanthine), flavonoids, tannins, saponins, terpenoids, coumarins, phenolics, polyphenols, and fatty acids (Naija et al. 2014; Ashraf et al. 2016; Saleem et al. 2021).

Potential biological activity: Antioxidant, hypolipidaemic, hypoglycemic, and antiophthalmic activities (Bennani-Kabchi et al. 1999; Saleem et al. 2021).

Toxicity: Cytotoxicity (Saleem et al. 2021).

Botanical description: Shrub up to 1.6(2) m in height and 2(3) m in diameter, very variable in growth form, mostly forming erect semi-globular bushes, more rarely prostrate mats, in living stage very variable in color, from glaucous to purplish and from dark to blackish green, drying black, youngest organs (preferably stems) covered with fugacious hairs; stem intricately branched from base, base up to 3 cm thick, young branches erect, ascending or spreading, pale green or purplish, smooth, with internodes of (1)2–5(8) mm, after shedding of leaves very rough by the remains of leaf bases, turning grey, finally longitudinally fissured; leaves succulent, (3)6–17(22) long, (1)1.5–4(8) mm wide, (0.7)1.2–2(2.5) mm thick, narrow leaves linear and terete or semi-terete, wider leaves linear or oblong, semi-terete or flat on both sides, broad leaves oblong to elliptic and even almost circular, flat on both sides, obtuse, at base attenuate into a petiole of up to 1 mm, or sessile, straight, ascending or spreading with C4 anatomy and central aqueous tissue; inflorescences terminal in simple or branched spikes, in upper parts mostly clearly flexuose; bracts (1.5)3–5(9) mm long, (1)2–2.5 mm wide, 1.5 mm thick, spreading or recurved, usually strongly arcuate, in case of dense floral clusters the upper sometimes hidden; flowers 3-very many in axillary clusters; bracteoles membranous, (0.5)0.7–0.9(1) × 0.3–0.7 mm, narrowly to broadly lanceolate, erose-dentate to ciliate; bisexual flowers ±drum-shaped, 1.0–1.5 mm wide, 0.5–0.6(0.8) mm high, female flowers 0.5–1 mm wide, ovoid to ellipsoid; tepals in bisexual flowers fused for about 1/3 and very succulent, in the female flowers shorter, fused for ½, non-succulent; stamens 5, inserted near tepal base; filaments (0.8)1–1.2(1.4) mm long, band-shaped, anthers (0.5)0.6–0.8(1) mm long; ovary ovoid to conical or pear-shaped, apically truncate, in female flowers often ±cylindric, (0.5)1–1.3 mm high, c. 0.5–0.6 mm in diam.; stigmas 3(4), (0.25)0.4–0.6 mm long, shortly papillose, inserted in the sunken center of the collar-like ovary apex; fruiting perianth in bisexual flowers moderately enlarged, 1.5–2.5 mm in diam., up to 1.5 mm high, at base conical, truncate, in female flowers ±unchanged; seeds in bisexual flowers horizontal, in female flowers vertical or oblique, 0.8–1.3 mm long, 0.65–1.0 mm wide, 0.5–0.7 mm thick, almost globular to drop-shaped, slightly flattened, beak short, testa black, shining, smooth (Fig. 2.40) (www.efloras.org).

Fig. 2.40 *Suaeda fruticosa* Forssk. ex J.F.Gmel.

Pollen features: Pollen type: apolar; pollen shape: spheroidal; pollen ornamentation: microechinate; polar diameter: 12.75–20.75 μm; equatorial diameter: 16.25–19.50 μm; exine thickness: 1.5–3.0 μm (Nazish et al. 2019).
Anatomical features: Amphistomatic leaves have been observed in the plant, with paracytic type of stomata (Nazish et al. 2020).
Part used: Leaves
Mode of utilization: Infusion, poultice
Route of administration: Oral, topical
Disease treated: Ophthalmia and emetic
Traditional uses: The plant is used as a diuretic, laxative, for pain, fever, skin, respiratory, toothache, digestive, and genitor urinary disorders (Saleem et al. 2021).
Recipes: Leaves of the plant are used as poultice applied to ophthalmia; infused in water and used as emetic (Malhotra et al. 1966; Saleem et al. 2021).
Potential economic significance: The data published by Wickens et al. (2012), Ozturk et al. (2019), Saleem et al. (2021), and Altay and Ozturk (2021) point out that the seeds and young shoots are consumed (cooked or raw) by humans. This species can be evaluated to reduce soil salinity and for remediation of hazardous metal-contaminated soils (Hameed et al. 2012; Caparrós et al. 2022). According to Ozturk et al. (2019), Altay and Ozturk (2021), and Saleem et al. (2021), the high salt content in the plants generally limits their use as stand-alone fodder crop but can be used as the component of a feed mix. In Pakistan, the plant is used as fodder for camels (www.efloras.org). This plant species is one among the number of sodium rich plants called "Barilla" which was used to make soda ash for soap production and evaluation in the glass industries. In the 18th and 19th centuries, India has exported large quantities, even now this plant together with various other chenopods are still collected from the seasonal salt marshes in the Rann of Kutch (India) for local use in the manufacture of soap and baking soda (Wickens et al. 2012).

2.33 *Tamarix aphylla* (L.) H.Karst.

Family: Tamaricaceae	
Synonym(s): *Tamarix articulata* Vahl; *Tamarix faras* Edgew.; *Tamarix furas* Buch.-Ham. ex Royle; *Tamarix orientalis* Forssk.; *Tamarix pharas* Buch.-Ham. ex Wall. (www.powo.science.kew.org).	
English name: Athel Tamarisk, Tamarisk Galls	
Flowering period: May–September	
Distribution: Afghanistan, India, Iran, Iraq, Kuwait, Oman, Pakistan, Palestine, Saudi Arabia, and Yemen (Malik et al. 2021).	
Altitude: Up to 1400 m	
Habit: Tree	
Ecological type: Xerohalophyte	
Habitat: High alkalinity-salinity soils	
Phytochemicals: Flavonoids, phenolics, tannins, steroids, cardiac glycosides, sesquiterpenes, monoterpenes, alkaloids, and essential oils (Alhourani et al. 2018; Alshehri et al. 2022).	
Potential biological activity: Analgesic, antioxidant, anti-inflammatory, antiproliferative, antidiabetic, antibacterial, antifungal, anticholinesterase, wound-healing and antipyretic activities (Alhourani et al. 2018; Alshehri et al. 2022).	
Toxicity: Cytotoxicity, and acute toxicity (Alhourani et al. 2018; Bahramsoltani et al. 2020).	
Botanical description: An evergreen tree to high shrub, slender, cylindrical, jointed branches articulate at the base of the sheath; the bark reddish-brown to grey and the slender twigs often hoary with deep punctate glands which produce a saline efflorescence; the salty "tears" drip in the night and the soil beneath trees is generally covered with a salt layer; the foliage fine bluish-grey or greyish-blue and superficially resembles long pine needles; the leaves are small and reduced to tiny scales that ensheathe the wiry twigs and are well equipped to withstand desiccation; the lamina reduced to a minute triangular tooth that marked with glands; the flowers loosely arranged on the slender spikes, small, usually white, occasionally pink, unisexual or bisexual, monoecious or dioecious, sessile and delicate, scattered on long, slender spikes usually clustered at the end of branches in loose racemose panicles, bracts sheathing; vernal inflorescences simple, aestival ones compound and more common; raceme 3–6 cm long, 4–5 mm broad, with sub-sessile flowers; bracts triangular to broadly triangular, acuminate, somewhat clasping, longer than pedicels; pedicel much shorter than calyx; calyx pentamerous; sepals ca. 1.5 mm long, entire, obtuse, the two outer slightly smaller, broadly ovate to broadly elliptic, slightly keeled, the inner slightly larger, broadly elliptic to sub-orbicular; corolla pentamerous, sub-persistent to caducous; petals 2–2.25 mm long, elliptic-oblong to ovate-elliptic; androecium haplostemonous, of five antesepalous stamens; filaments inserted between the lobes of the nectary disc; fruit small, bell-shaped sessile capsule, and ripens in the cold season; the capsules contain minute seeds with tufts of long, soft, woolly hairs; ripe capsules turn brown and open up gradually to allow the minute seeds to be dispersed by the wind (Fig. 2.41) (Baum 1978; National Academy of Sciences 1980; www.cabi.org).	
Pollen features: Pollen shape: prolate, surface ornamentasyon: reticulate, polar axis: 13.30–18.56, equatorial diameter: 10.11–12.77 (Elkordy and Faried 2017).	

Fig. 2.41 *Tamarix aphylla* (L.) H.Karst.

Anatomical features: Roots: Epidermal layer consists of closely elongated cells. Thick parenchymatous cortex. Vascular bundles are arranged radially, protoxylem towards the center while metaxylem towards the cortex. Pith, clear (Bibi et al. 2015).
Stem: Thick epidermis. Cortex is thick and composed of parenchymatous cells. Vascular bundles are radially arranged from the center towards the periphery. Pith large and clear (Bibi et al. 2015).
Leaf: The epidermis is composed of elongated cells (Fig. 2.42). It covers the cortex composed of mesophyll tissues. Water storing tissues are many in number. Vascular bundles, in rings (Bibi et al. 2015).

Part used: Whole plant

Mode of utilization: Smoke, infusion, decoction, powder, paste, extract, and juice

Route of administration: Oral, topical, and inhale

Disease treated: Hepatitis, stomachic, leukorrhea, sexual weakness, jaundice, menstrual problems, cooling agent, eczema, wound healing, burnt, chicken pox, ringworm, abscess, earache, aphrodisiac, tonic, diuretic, throat infections, obesity, and spleen diseases.

Traditional uses: The leaves are used for wounds and abscess healing, as astringent, and for rheumatism and joint pain (Alhourani et al. 2018); also used as carminative, eczema, antimicrobial, antioxidant, aphrodisiac, anthelmintic, diuretic, anti-hemorrhoid, antidiarrheal, and skin diseases (Alshehri et al. 2022). The plant is used to treat various infectious diseases, including dental infections, smallpox, leprosy, tuberculosis, colds, and coughs (Alshehri et al. 2022).

Recipes:
- Smoke of whole plant is used to treat fever.
- Decoction of stem bark is used to take bath to treat skin allergies.
- Sweet substance on twigs is mixed in cold water to make a juice which is taken as cooling agent.
- Bark infusion is mixed with butter oil and applied dermally for wound healing.
- Wood powder is applied dermally on burnt skin to reduce pain.
- Twigs paste is applied dermally to treat chicken pox.
- Decoction of young stem and leaf is used as lotion to cure spleen swelling.
- Extract of flower galls is used as gargle to treat throat infections.

Potential economic significance: The rural carpentry workers in Pakistan and India are evaluating the wood and also commonly using it as fuel source (Khan et al. 2005).
It has good potential if used for the stabilization and afforestation of sand dunes, in shelterbelts and as windbreaks (www.cabi.org). It has also been introduced in Australia as a wind breaking hedge plant, and as an ornamental tree (Griffin et al. 1989; www.powo.science.kew.org).

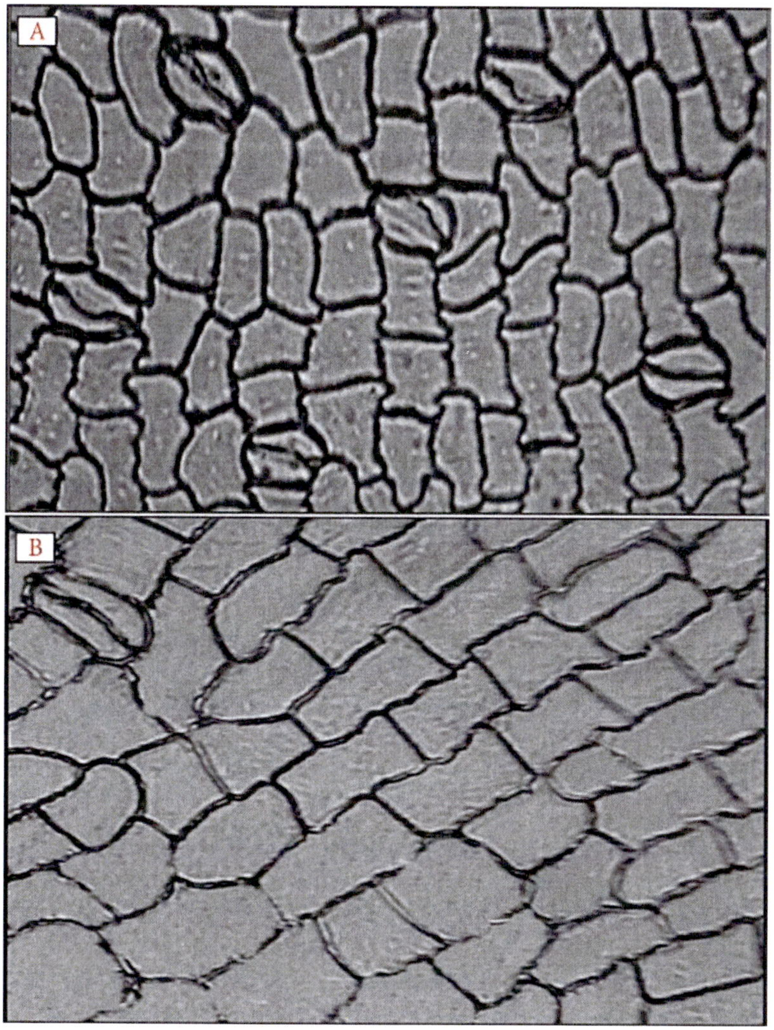

Fig. 2.42 (a) Abaxial surface; (b) Adaxial surface of *Tamarix aphylla* (L.) H.Karst.

2.34 *Tamarix dioica* Roxb. ex Roth

Family: Tamaricaceae
Synonym(s): *Tamarix articulata* Wall., *Tamarix longe-pedunculata* Blatt. & Hallb.
English name: Saltcedar, Red Tamarisk, Tamarisk
Flowering period: April–November
Distribution: Afghanistan, Bangladesh, East Himalayas, India, Iran, Myanmar, Nepal, and Pakistan (Malik et al. 2021).
Altitude: 300–1500 m
Habit: Shrub
Ecological type: Xerophyte
Habitat: Saline soils and dry places
Phytochemicals: Flavonoids, saponins, terpenoids, steroids, tannins, phenols, pholbatannins, amino acids and essential oils (Bughio et al. 2017).
Potential biological activity: Antifungal, antibacterial, ulcer-protective and anti-inflammatory, hepatoprotective, antioxidant/free radical scavenger activities, cytoprotective against gastric epithelial cell damage, and cardioprotective potential (Ahmad et al. 2002; Najmi et al. 2004; Ashour et al. 2012; Zaidi et al. 2012; Nidavani et al. 2014; Bughio et al. 2017; Imtiaz et al. 2019).
Toxicity: Hepatotoxicity (Bahramsoltani et al. 2020).
Botanical description: Shrub or under-sized tree with reddish bark, entirely glabrous, leaves vaginate, abruptly acuminate, 1.5–3 mm long, with a broad whitish margin; dioecious plants; recemes aestival, simple or loosely compound male racemes, 3–8 cm long, 7–8 mm broad, compact, female raceme as long as the male raceme or somewhat longer, 3–5 mm broad; flowers purple or purplish-pink, subsessile; bracts semiamplexicaul, triangular, acuminate, 2.5–3 mm long, 1 mm broad; sepals 5, ovate to broadly ovate or somewhat orbicular, c. 2 mm long, c. 1.5 mm broad, almost entire, obtuse; petals 5, free, obovate, 2.5 mm long, 1–1.25 mm broad, entire, obtuse; stamens 5, filaments filiform, 1.75–2 mm long, inserted in between the notches of the disc, notches almost entire to slightly emarginate, anthers obtuse, ± sagittate; stamens abortive in the female flowers; styles 3 as long as the ovary or slightly shorter, exserted, ovary triquetrous 1–1.5 mm long, abortive or absent in male flowers (Fig. 2.43) (www.efloras.org).
Pollen features: Pollen shape: prolate-spheroidal, polar length: 19.36 μm, equatorial breadth: 16.9 μm (Qaiser and Perveen 2004).

Fig. 2.43 *Tamarix dioica* Roxb. ex Roth

Anatomical features: We find epidermal cells with little more thickened inner walls than outer walls; pits are not deep; cortical parenchyma made up of small cells; a sclerenchymatous pericycle more or less forms a loose ring of stone cells with thickened and radially striated interposing cells; the wood forms a composite hollow cylinder; medullary rays are 2–6 seriate; pith is composed of thick-walled cells; pericycle made up of large groups of stone cells; the stone-cell groups in the plant are closely placed all around the soft bast and are interposed between them; being characterized by sclerosis and radical striation of the wall; wood in the plant forms a composite hollow central cylinder enclosing a small pith tissue, consisting of xylem bundles connected together by interfascicular wood prosenchyma; medullary rays are 2–6 seriate; pith is formed of thick-walled cells, those near the periphery being filled with granules (Sabnis 1921).

Part used: Stem, root, bark, seed, flower, leaf

Mode of utilization: Decoction, powder, paste, raw

Route of administration: Oral and topical

Disease treated: Kidney stones, hepatitis, leprosy, spleen tumor, edema, dysentery, leukorrhea, menstrual problems, aphrodisiac, syphilis, ulcer, sores, hair lice, burns, toothache, measles, blisters, backache, anemia, stop vomiting, and snakebite.

Traditional uses: The leaves are used to cure liver infection, splenic or inflammation; as a diuretic and carminative. According to Said (1984), Zaidi et al. (2012), Bughio et al. (2017) and Imtiaz et al. (2019), the plant is used in the case of indications such as vaginal discharge, in the treatment of diarrhea, dysentery and inflammation, also used for colds, fever, flu, cough and as an astringent in burns, being leukodermic and styptic in nature. The reports published by Anis et al. (2000), Shah and Hussain (2012), Waaris et al. (2014), Chaudhary and Kumar (2015) and Imtiaz et al. (2019), a paste of the dried bark and leaves is reported to have a soothing effect on wounds, the plant is active against various serious bacterial and viral infections such as tuberculosis, leprosy, gonorrhea, ringworm, polio, and measles.

Recipes:

- Twigs boiled with wine taken to treat spleen tumor.
- Powder of root bark is taken as aphrodisiac.
- Paste of twigs bark is applied externally to treat snake bite venom.
- Seed decoction with wine and honey is applied externally to treat chest sores.
- Flower decoction is used to wash hair to kill hair lice.
- Young stem is chewed to treat toothache.
- Leaf decoction with *Arctium lappa* seeds is applied externally to treat measles.
- Leaf ash with honey is applied externally to treat blisters.

Potential economic significance: The plant is cultivated as ornamental and is also used in very low quantities for fuelwood (Shah et al. 2014). It has also been locally used as fodder plant (Kundu 2009).

In the flooded lands along the rivers, the plant springs up in the form of considerable thickets and is used for wattling, baskets and roofs by local people (Yousaf 2014). The branches are also used as brooms (Rajesh et al. 2013).

2.35 *Tribulus terrestris* L.

Family: Zygophyllaceae	
Synonym(s): *Tribulus lanuginosus* L.; *Tribulus saharae* A. Chev.; *Tribulus terrestris* var. *sericeus* Andersson ex Svenson (www.powo.science.kew.org).	
English name: Devil's thorn, Puncture vine	
Flowering period: Almost throughout the year (www.efloras.org; Malik et al. 2021).	
Distribution: Afghanistan, Bangladesh, Cambodia, China, East Himalayas, India, Iran, Iraq, Kazakhstan, Kyrgyzstan, Kuwait, Lebanon, Mongolia, Myanmar, Nepal, North Caucasus, Oman, Pakistan, Palestine, Russia, Saudi Arabia, Sri Lanka, Syria, Tadzhikistan, Taiwan, Thailand, Tibet, Transcaucasus, Turkiye, Turkmenistan, Uzbekistan, Vietnam, West Himalayas, West Siberia, Yemen (Malik et al. 2021).	
Altitude: Up to 3500 m	
Habit: Herb	
Ecological type: Xerophyte	
Habitat: In dry, loose, sandy soils and prospers near sand dunes or loose blown soil around field margins (Holm et al. 1977).	
Phytochemicals: Flavonoids, flavonol glycosides, steroidal saponins, and alkaloids (Chhatre et al. 2014).	
Potential biological activity: Diuretic, aphrodisiac, antiurolithic, immunomodulatory, antidiabetic, absorption enhancing, hypolipidemic, cardiotonic, central nervous system, hepatoprotective, anti-inflammatory, analgesic, antispasmodic, anticancer, antibacterial, anthelmintic, larvicidal, and anticariogenic activities (Chhatre et al. 2014).	
Toxicity: Hepatotoxicity, nephrotoxicity and neurotoxicity (Talasaz et al. 2010), and cytotoxicity (Chhatre et al. 2014).	
Botanical description: Annual or biennial, prostrate, densely appressed whitish silky pubescent herb; stem hirsute to sericeous, branches spreading; leaves paripinnate, 2.5–5 cm long; stipules lanceolate to falcate, 3–5 mm long; leaflets (4-) 5–6(−8) pairs ovate to elliptic-oblong, 5–10(−12) mm long, 3–8 mm broad, inequilateral, acute; flowers yellow, 1–1.5 cm across; pedical up to 1.5(−2) cm long; sepals ovate-lanceolate, 5–6 mm long, c. 3 mm broad, acute; petals obovate, 6–8 mm long, 3–4 mm broad, obtuse; stamens 10, filaments c. 3–5 mm long, anthers versatile; ovary ovoid, hirsute; style c. 1.5 mm long, stigmas decurrent; fruit up to c. 1 cm in diameter, 4–8 mm long, mericarps densely crested and tuberculate on dorsal side, densely hairy to glabrescent, with 2 long patent and 2 short downwardly directed spines (Fig. 2.44) (www.efloras.org).	
Pollen features: Pollen grains: pantoporate; pollen shape: oblate-spheroidal; polar length: 32–52 μm, equatorial length: 36–48 μm; exine ornamentation: reticulate (Perveen and Qaiser 2006; Semerdjieva et al. 2011).	
Anatomical features: A C4 plant with anatomical structural adaptations typical in this connection as per Nikolova and Vassilev (2011). It shows typical distinguished xeromorphic adaptations; lamina is isobilateral and amphistomatous, with a thick cuticle; the "Kranz" anatomical structure, mesophyll with reduced intercellular spaces, small stomata in the epidermis, very low stomatal density with many unicellular trichomes, particularly on the lower epidermis. According to Nikolova and Vassilev (2011), its poor performance under shade and limitations in the distribution are explainable with its leaf structure; stem and root have compact structures with well-developed vascular tissue, high sclerification and little parenchyma.	
Part used: Fruit	
Mode of utilization: Powder	
Route of administration: Oral	
Disease treated: Urinary disorders, and to cure impotency	

Fig. 2.44 *Tribulus terrestris* L.

Traditional uses: In India, the plant is used as a tonic and diuretic, and for the treatment of painful urination, calculous affections and Bright's disease (www.cabi.org).
The fruits have been used as diuretic and cough expectorant as reported in the Chinese Pharmacopoeia. It is reported to improve eyesight and is reported to be good for treatment of skin pruritus, headache and vertigo, as well as mammary duct blockage (Zhu et al. 2017). In India in Ayurvedic system, the fruits are used for the treatment of infertility, impotence, erectile dysfunction and low libido. It has been used in Pakistan as diuretic and has uricosuric effects (Akram et al. 2011; Mohammed et al. 2014; Zhu et al. 2017).

Recipes:

- Fruit powder is given orally to cure urinary disorders and mixed with sugar to ease delivery (Qureshi et al. 2009).
- The powder of fruits is taken orally with a glass of milk by rural males to cure impotency (Qureshi et al. 2009).

Potential economic significance: In India, this plant is cultivated to bring down and protect soil erosion resulting from both wind and water. It also reported to prove helpful in the protection of loss of soil moisture; also used to improve soil texture and water holding capacity in wastelands (Brajeshwar 2001–2002; www.cabi.org).

2.36 *Typha angustifolia* L.

Family: Typhaceae

Synonym(s): *Massula angustifolia* (L.) Dulac; *Typha angustifolia* subsp. *angustata* (Bory & Chaub.) Briq.; *Typha elatior* Boenn.; *Typha foveolata* Pobed.; *Typha glauca* Seg.-Vianna; *Typha gracilis* Rchb.; *Typha media* C.C.Gmel.; *Typha minor* Curtis; *Typha pontica* Klokov f. & Krasnova (Malik et al. 2021).

English name: Lesser bulrush, Narrowleaf cattail, Lesser reedmace (Malik et al. 2021).

Flowering period: June–September (Malik et al. 2021)

Distribution: China, India, Kazakhstan, Kyrgyzstan, Lebanon, Nepal, Pakistan, Palestina, Russia, Syria, Taiwan, Thailand, Turkiye, Turkmenistan, Uzbekistan, Vietnam (Malik et al. 2021).

Altitude: Up to 1930 m

Habit: Herb

Ecological type: Hydrohalophyte

Habitat: In the shallow lake waters, alongside the rivers, in ponds, marshes, and ditches (Malik et al. 2021).

Phytochemicals: Alkaloid, protein, glycoside, tannins, steroid, phenols, saponin, flavonoid, carbohydrate, oil and fat (Londonkar et al. 2013; Malik et al. 2021).

Potential biological activity: Immunosuppression, antiplatelet aggregation, antimicrobial, cholesterol-lowering and antithrombolytic activities, and antiatherogenic effect (Umesh et al. 2014).

Toxicity: Heavy metal toxicity (Demirezen and Aksoy 2004), and cytotoxicity (Umesh et al. 2014).

Botanical description: Herbaceous, stout, stem is 1.6–3.0 m tall; leaves 53–121 cm × 4.0–9.0 mm, lower side is convex, transversal sections semicircular; male spike 8.2 cm long, with 1.0–3.0 deciduous bracts; female spike (5.0-) 15.1–30.0 cm long, separated from the male by an axis 2.6–7.0 cm; male flowers have 4 stamens, rarely 3 or 5; anthers 2.0 mm; female flowers with bracteole; bracteoles fusiform; ovary too fusiform; stalk 5.0 mm, cylinderical; style 1.0–1.4 mm; stigma lanceolate to linear, 1.4–1.9 mm; the hair on the stalk shorter than style; fruit narrow elliptical (Malik et al. 2021) (Fig. 2.45).

Fig. 2.45 *Typha angustifolia* L.

Pollen features: Pollen: monads, exine ornamentation: reticulate, polar length: 9.40–15.93 μm, equatorial length: 14.65–24.17 μm (Hamdi et al. 2010).

Anatomical features: A wedge-shaped lamina margin, having thick zone of fibers at the margin, containing 1–4 vascular bundles, embedded at the proximal edge of the zone (Mcmanus et al. 2002). The abaxial and adaxial margins of the leaf are studded with subepidermal vascular bundles, interspersed with fiber bundles in the chlorophyllous mesophyll. According to Mcmanus et al. (2002), the epidermis located above the vascular bundles lacks enlarged epidermal cells and extra thickened cuticles resulting in the surface being smooth to the touch.

Part used: Leaves, flowers

Mode of utilization: Powder, decoction

Route of administration: Oral, topical

Disease treated: Gynecological disorders

Traditional uses: Leaves are used to relieve abdominal pains, as carminative- antispasmodic, lessen gastric acidity as an antiseptic and antibiotic. Its use as antilice and antidandruff agent is reported in the form of water extract of leaves. The flower is a carminative, used in digestive disorders and for colic pains (Malik et al. 2021).

Recipes: (Malik et al. 2021)

- Dry leaves crushed into fine powder (40 g), 2 tablespoons used with 2-glasses (200 mL) of water for 18 days, 2 times a day daily for treating genital tract infections.
- Flowers (70 g) boiled with 200 mL of water for 9 min, infusion filtered, used regularly for 20 days for treating gynecological disorders.

Potential economic significance: *Typha* plants play an important role in wetland ecology. Dvorak (1996) has reported that these plants provide food, shelter and nesting sites for waterfowl, fish and other wildlife; also playing an important role in wetland restoration projects (Dobberteen and Nickerson 1991). The species has a great significance in navigable river systems for protecting banks from erosion (Bonham 1980).

In the past, paper has also been made from *Typha* pulp. According to Holm et al. (1991), a set of books published in 1765 with *Typha* paper still available (www.cabi.org). This plant has also been considered as a biomass crop for energy purposes (Yurukova and Kochev 1993). It can serve as a valuable component of constructed wetlands, for wastewater cleanup, as they are efficient in the accumulation of heavy metals, and in removing nutrients and oils (Tjitrosoedirdjo and Sastroutomo 1986; www.cabi.org).

2.37 *Typha latifolia* L.

Family: Typhaceae
Synonym(s): *Massula latifolia* (L.) Dulac; *Typha ambigua* Schur ex Rohrb.; *Typha crassa* Raf.; *Typha elata* Boreau; *Typha elatior* Boreau; *Typha elatior* Raf.; *Typha elongata* Pauquy; *Typha elongata* (Dudley) Dudley; *Typha engelmannii* A.Br. ex Rohrb.; *Typha intermedia* Schur; *Typha latifolia* f. *divisa* Louis-Marie; *Typha latifolia* var. *elongata* Dudley; *Typha latifolia* subsp. *eulatifolia* Graebn.; *Typha latifolia* var. *obconica* Tkachik; *Typha latifolia* var. *typica* Rothm.; *Typha major* Curtis; *Typha media* Pollini; *Typha palustris* Bubani; *Typha pendula* Fisch. ex Sond.; *Typha remotiuscula* Schur; *Typha spathulifolia* Kronf. (www.powo.science.kew.org).
English name: Broadleaf cattail bulrush, common bulrush, common cattail
Flowering period: June–August
Distribution: Afghanistan, China, Georgia, India, Kazakhstan, Kyrgyzstan, Lebanon, North Caucasus, Pakistan, Palestine, Russia, Syria, Tadzhikistan, Tibet, Transcaucasus, Turkiye, Turkmenistan, Uzbekistan, and West Siberia (Malik et al. 2021).
Altitude: Up to 2300 m
Habit: Herb
Ecological type: Hydrohalophyte
Habitat: Marshes, wet meadows, lakeshores, roadside ditches, seacoast estuaries, pond margins, and fens (Grace and Harrison 1986).
Phytochemicals: Saponins, phenols, alkanolamides, flavonol glycosides, allelopathic sterols, alkaloids, and fatty acids (Gescher and Deters 2011; Shinwari et al. 2019).
Potential biological activity: Antimicrobial activity (Shinwari et al. 2019).
Toxicity: Phytotoxicity (Aliotta et al. 1990).
Botanical description: Plants up to 100–200 cm high; stem thick, terete; leaves linear-broadly linear, 8–20 mm broad; flowering stem equal to or somewhat shorter than the leaves; male and female parts of inflorescence contiguous; female parts slightly longer than the male parts at maturity, cylindrical, soft, dark brown or blackish brown; male flowers with simple hairs and pollen in tetrads; filaments 2–3 times as long as anther; female flower ebracteate, ovary 1/3–1/4 the length of stipe, stigma lanceolate or rhombic, fleshy, dark brown or persistent brown, much surpassing the perianth hairs (Fig. 2.46) (www.efloras.org).
Pollen features: Pollen: tetrads, exine ornamentation: perforate-reticulate, polar length: 23.21–27.80 µm, equatorial length: 25.20–34.22 µm (Hamdi et al. 2010).
Anatomical features: The leaf lamina margin oblong, often curved in shape, with a zone of fibers at the margin, one vascular bundle embedded within the proximal curved zone of fibers. The abaxial and adaxial margins of the leaf show that the subepidermal vascular bundles are interspersed with fiber bundles in the chlorophyllous mesophyll; epidermal cells are located above each vascular/fiber bundle and enlarged, thickly cutinized; giving the surface of the leaf a ridged or ribbed effect (Mcmanus et al. 2002).
Part used: Rhizome, roots
Mode of utilization: Pounded rhizome, roots
Route of administration: Topical
Disease treated: Soothing of burns

Fig. 2.46 *Typha latifolia* L.

Traditional uses: In Pakistan, the root is used as an aromatic compound tonic for brain and heart; demulcent during fever, anti-diuretic especially in children and to clot blood (Ali et al. 2020).
Leaves and pollens are also used as an astringent, diuretic, sedative and anticoagulants, also in the treatment of kidney stones and painful menstruation (Ismail and Nisar 2010).

Recipes: Pounded roots of the plant are used for the soothing of burns (Iqbal et al. 2019).

Potential economic significance: *Typha* plants are an important component in the wetland ecology. They provide food, shelter and nesting sites for huge number of water birds (Dvorak 1996), also help in the wetland restoration (Dobberteen and Nickerson 1991). The species is valuable for erosion control alongside the banks in navigable river systems (Bonham 1980). The paper has been produced with *Typha* pulp in the past. Holm et al. (1991) state that some books published in 1765 with *Typha* paper (www.cabi.org). It has also been considered as a biomass crop for energy purposes (Yurukova and Kochev 1993).
A valuable component of constructed wetlands, used for wastewater cleanup, efficient at accumulating heavy metals, as well as in removing nutrients and oils (Tjitrosoedirdjo and Sastroutomo 1986; www.cabi.org). The plant is also cultivated as an ornamental (www.cabi.org).

2.38 *Vachellia nilotica* (L.) P.J.H.Hurter & Mabb.

Family: Fabaceae
Synonym(s): *Acacia nilotica* (L.) Willd. ex Delile; *Gumifera nilotica* (L.) Raf.; *Mimosa nilotica* L.
English name: Gum Arabic Tree
Flowering period: August–October
Distribution: Bangladesh, East Himalayas, India, Iran, Myanmar, Nepal, Oman, Pakistan, Saudi Arabia, Sri Lanka, West Himalayas, Yemen
Altitude: Up to 1500 m
Habit: Tree
Ecological type: Xerophyte
Habitat: The plant is most commonly found growing in grasslands, pastures and open woodlands; near water sources in arid areas (www.keyserver.lucidcentral.org).
Phytochemicals: Alkaloids, flavonoids, polyphenols, tannins, terpenes, sitosterol and fatty acids (Rather et al. 2015).
Potential biological activity: The plant possesses remarkable therapeutic potential against various diseases such as bacterial, fungal, viral, amebic, leukodermal diseases, bleeding piles, hypertension, hemorrhoid, cancer, congestion, menstrual problems and earache (Rather et al. 2015).
Toxicity: Cytotoxic (Rather et al. 2015).
Botanical description: Tree, 1.2–18 m high, variable in shape; bark on trunk rough, fissured, blackish, grey or brown; young branches almost glabrous to subtomentose; stipules spinescent, up to 8 cm long; leaf often with 1–2 petiolar glands and others between all or only the topmost of the 2–11 pairs of pinnae; leaflets 7–25 pairs, 1.5–7 mm long, 0.5–1.5 mm wide, glabrous to pubescent; inflorescence axillary pedunculate heads, 6–15 mm in diameter; flowers bright yellow, calyx 1–2 mm long, pubescent or subglabrous; corolla 2.5–3.5 mm long, glabrous to more or less pubescent outside; fruit very variable, indehiscent, straight or curved, glabrous to velvety, 4–22 cm long, 1.3–2.2 cm wide; seed blackish brown, smooth, 7–9 mm long, 6–7 mm wide, subcircular, compressed, areole 6–7 mm long, 4.5–5 mm wide (Fig. 2.47) (www.tropicos.org).

Fig. 2.47 *Vachellia nilotica* (L.) P.J.H.Hurter & Mabb.

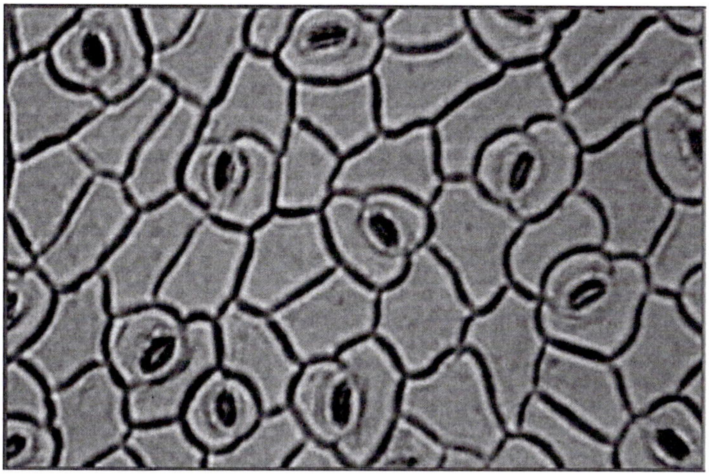

Fig. 2.48 Abaxial surface of *Vachellia nilotica* (L.) P.J.H.Hurter & Mabb.

Pollen features: Pollen shape: rounded; pollen size - polar axis: 50 μm, equatorial diameter: 48 μm; type of colpi: colporate; tectum surface: psilate-foveolate (Al-Watban et al. 2013).

Anatomical features: Majority of the fusiform initials are multinucleate in the non-stratified cambium of this tree; length increases down the stem from the apex, attaining a maximum in the old trunk and declining slightly near the base; short and uni- to biseriate rays are most abundant in the young shoots, whereas tall and multiseriate rays dominate the cambial surface in the trunk region throughout the year. The population is minimum in the early phase of cambial activity, maximum during peak activity; overall proportion of fusiform initials in the cambial cylinder initially increases with age, from young shoots towards the base, later becomes more or less constant in the trunk region (Iqbal and Ghouse 1987).

There are distinct growth rings, marked by defined growth zones and thin banded parenchyma; vessels are diffused, solitary and radial; as multiples of 2–4, perforations simple; inter vessel pits vestured, vessel-ray and vessel parenchyma pits same in type and size to intervessel pits; tyloses and substances rarely found in the vessels; sparse to moderately abundant paratracheal, vascicentric parenchyma, usually in prominent sheaths, 2–4 cells wide around the pores, tending to form particularly around the smaller pores in 2–4 celled strands; the rays are 1–4 seriate; fibers are without septate; material inclusion as prismatic crystals found in chambered parenchyma strands (Krisdianto and Damayanti 2007).

Epidermis present above paranchymatous mesophyll tissues in the leaves (Fig. 2.48); vascular bundles clear, in the ring form around the vascular tissues; bundle sheath cells present, the characteristic feature of the xerophytes; epidermis encloses the cortex. In the stem; cortex is composed of collenchymatous cells; beneath it the endodermis is found, next to the pericycle, radially arranged parenchymatous tissues present with the capacity to retain large quantity of moisture; numerous vascular bundles in the center and some are in scattered form; epidermis surrounding the cortex in the roots; with scattered vascular bundles, protoxylems in the center while metaxylems towards the cortex, xylem vessels, many in number, the plant a typical xerophyte (Bibi et al. 2014).

Part used: Bark, leaf, fruit, flower, stem, gum

Mode of utilization: Decoction, infusion, extract, powder

Route of administration: Oral, topical

Disease treated: Mouth sores, teeth and gum disorders, blood purifier, diabetes, diabetes, asthma, aphrodisiac, liver diseases, arthritis, eye diseases, high blood sugar, impotency, leukorrhea, night fall, sexual debility, hepatitis, and heart tonic.

Traditional uses: All parts of the plant have been used as gastrointestinal disorders, diarrhea, aphrodisiac, dressing of ulcers, anti-inflammatory, in Alzheimer's disease, wounds, pharyngitis, bronchitis, and diabetes, in cancers (especially ear, eye, and testicles), tuberculosis, and indurations of liver and spleen (Rather et al. 2015).

Recipes:

- Stem bark is soaked in tepid water for few hours; water used for mouth sores and teeth and gum disorder.
- Fresh leaves are boiled in water and orally used for straining twice in day for dysentery.
- Flower extract is used to treat women infertility.
- Fruit infusion is used to treat hepatitis.
- Leaf decoction is used to treat cough.
- Gum mixed in sugar and wheat flour is roasted in oil and used as heart tonic.
- The powder of young dried branches is used to treat sexual debility.

Potential economic significance: The gum extracted from the pod is used for making inks and dyes. Wood is very hard, durable and preferred for agricultural implements. Fruits are used as cattle feed (www.indiabiodiversity.org).

The plant gum is known as samogh or samuk (Arabic). It is sold in balls, being a commercial of inferior quality; used as an emulsifying agent and emollient; is edible and used to relieve throat and chest complaints (www.powo.science.kew.org).

The pods are desirable fodder for cattle; the leaves, young shoots and young pods are reported to be useful in the milk production (www.powo.science.kew.org); wood burns without too much smoke and provides good charcoal; flowers are attractive for bees as nectar source and provide pollen for them. The plant gives suitable product for wheels, railway sleepers, boat building, mine timber, live fencing and water wells; its wood is durable and resistant to borers and termites. In ancient Egypt, the sap- and heart-wood have been used for furniture, paneling, house beams and statues, it was regarded as impervious to insect and fungus attacks. In India, this tree is used as an ideal windbreak used to encircle the fields (www.powo.science.kew.org). Adhikesavan et al. (2015) have reported that biodiesel is produced from seed oil of the plant.

The tree possesses dark brown heartwood, clearly distinct from reddish brown color of sapwood, the denser cell wall is attractively streaked in tangential surfaces; length of wood fiber decreases from pith toward periphery; longitudinally, higher stem has shorter fiber; wood of the tree has second class quality of fiber, which is moderately thick with narrow lumen diameter; timber is suitable for carved products. Its timber is suitable for charcoal production and fuel wood because of its high calorific value (Krisdianto and Damayanti 2007).

The tree holds great importance for reforestation, reclamation of wastelands and soil improvement; moreover micro propagation and symbiotic relationship with rhizobium and mycorrhizal fungi makes it one of the important species for increasing soil fertility (Rather et al. 2015).

2.39 *Withania somnifera* (L.) Dunal

Family: Solanaceae

Synonym(s): *Physalis somnifera* L.; *Withania kansuensis* Kuang & A. M. Lu; *Withania microphysalis* Suess. (Malik et al. 2021).

English name: Ashwagandha, Indian ginseng, poison gooseberry, winter cherry (Malik et al. 2021).

Flowering period: October–June (Malik et al. 2021).

Distribution: West and South Asia, Pakistan, India, and Sri Lanka (Kapoor 1990; Kirtikar et al. 1993; Malik et al. 2021).

Altitude: Up to 1900 m (Malik et al. 2021).

Habit: Shrub

Ecological type: Xerophyte

Habitat: Grows in dry areas in the Himalayas, open places, disturbed areas, and as under shrubs in stony places; grows as a weed along roadsides and in open waste places (Malik et al. 2021).

Phytochemicals: Phenolics, withanolide A, 12-deoxywithastromonolide and withaferins A, alkaloids, steroids, volatile oils, cysteines, glycosides, hentriacontanes, dulcitosl, withaniols, prolines, alanines, tyrosines, glycines, hydroxyprolines, aspartic acids, glutamic acids, cystines, valines and starchs (Dhanani et al. 2017; Malik et al. 2021).

Potential biological activity: The reports from Umadevi et al. (2012) and Malik et al. (2021) enlighten the facts that this plant is an excellent rejuvenator, a general health tonic, and a cure for a number of health complaints; diuretic, anti-inflammatory, a sedative and generally respected for increasing energy, endurance, acting as an-adaptogen, a strong immunostimulatory and anti-stress agent; taken for treating Parkinson's disease, bronchitis, asthma, impotence, nervous disorders, diabetes, conjunctivitis, epilepsy, ulcers, emaciation, insomnia, senile dementia, leprosy, rheumatism, arthritis, intestinal infections, cold and coughs, and a suppressant in HIV/AIDS patients.

Toxicity: Shows subacute toxicity and copper toxicity (Aphale et al. 1998; Khatun et al. 2008; Malik et al. 2021).

Botanical description: Small or medium undershrub or attain the height of a small tree, ascending and spreading, branching, 30–150 cm, at places 1–2 m in height, erect, grayish, perennial, with strong disagreeable odor like horse's urine, evergreen, cultivated all over the world, plant covered with minute-stellate tomentum; the plant is covered and surrounded with very short, small, fine, branched hairs and silver-grey in color; roots of Ashwagandha are fleshy when dry, straight, cylindrical, tapering down, gradually unbranched 10–17.5 cm long and 6–12 mm diameter in thickness; main roots are stout, fleshy and whitish brown in color, brownish outer and creamy interior, have fiber similar to the secondary roots having acrid taste (Naveen et al. 2015); branches ligneous, tomentose at the apex; leaves simple, 2–6 cm wide, 3–8 cm long, alternate, petiole 1–2 cm long, ovate, glabrous, simple over 10 cm long, simple, petiolate, elliptic-ovate to broadly ovate, entire, exstipulate, cunate or oblique, glabrous, sometimes up to 10 cm long, those in the floral region are smaller and opposite, arranged alternately (opposite on flowering shoots), simple, margins slightly waved, narrowing into 5–20 mm long petioles, broadly ovate or oblong, 29–80 mm long and 21–50 mm broad, generally stellate-tomentose, grayish; flowers generally small, greenish, axillary, monoceous or bisexual and solitary or in few-flowered cymes; corolla is 5-lobed, constrictly campanulate, 5–8 mm long and light yellow to yellow-green in color; fruit is usually a round hairless berry, 5–8 mm across, orange-red to red in ripped condition and is enveloped by the enlarged calyx; many seeds, discoid, reniform and yellow, most of the seeds are very pale brown, 2.5 mm across, sometimes kidney-shaped and squeezed with a rough surface and netted surface; seeds poisonous; the red fruit is enclosed by the brownish, papery, turgid calyx; a bad-smelling bush with generally strong-smelling roots, leaves are mentioned as having strong smell of green tomatoes (Kapoor 1990; Kirtikar et al. 1993; Mirjalili et al. 2009; Naveen et al. 2015; Malik et al. 2021) (Fig. 2.49).

Fig. 2.49 *Withania somnifera* (L.) Dunal

Pollen features: The pollen grains are tri- or tetrazonocolpate, measuring 29 μm. Surface sculpturing of the pollen is scarbate-granulate (Alwadie 2002).
Anatomical features: The leaves contain 4 morphologically distinct trichome types: glandular capitate, non-glandular dendritic (branched), non-glandular bicellular, and non-glandular multicellular; cuticular striations visible on both foliar surfaces, radiating from the bases of capitate and dendritic trichomes as well as from stomata; cuticular striations more frequent on abaxial surfaces, decreasing with increasing leaf development; stomatal guard cells surrounded by 3 subsidiary cells of differing sizes, arrangement of stomata on adaxial and abaxial surfaces of the plant leaves is anisocytic (Fig. 2.50) (Munien et al. 2015).
Part used: Leaf, root, fruit, bark, and stem
Mode of utilization: Decoction, powder, paste, extract, infusion, and raw
Route of administration: Oral and topical
Disease treated: Rheumatism, aphrodisiac, kidney stones, phlegm, constipation, improves memory, scorpion sting, hypertension, fever, analgesic, leukoderma, arthritis, cancer, asthma, stress, paralysis, body selling, and old age debility.
Traditional uses: In traditional medicine, the leaves as well as roots are used, both in external applications and as tonics (Senthil et al. 2009; Chatterjee et al. 2010; Singh et al. 2010a; Munien et al. 2015). The plant concoctions act upon the nervous and reproductive systems, with rejuvenating effects on the whole body (Chatterjee et al. 2010; Ram et al. 2012). It is widely used to improve vitality, as well as help in the recovery of various illnesses (Khan et al. 2006; Chatterjee et al. 2010; Kumar and Kumar 2011; Ram et al. 2012; Munien et al. 2015).
Recipes: • Fresh leaves are boiled in water; one cup of decoction is prescribed to take once in a day for rheumatism. • Root powder is given with milk to remove kidney stones and phlegm. • Root paste is applied externally to treat swelling. • Root extract is used to cure debility in old age people. • Fruit decoction is diuretic. • Stem bark infusion is taken internally to treat asthma. • Young shoots are eaten as vegetable to reduce obesity in results.
Potential economic significance: An important plant both ecologically and economically. It is as old as Ayurvedic medicine, with millennia of usage as one of the essential plants used in Ayurvedic medicine. The Mediterranean region, and several Oriental countries are using it (Afewerky et al. 2021). In India, it is readily available via cultivation and in the wild agricultural land; and aside from being medicinal, it is also used for bioremediation, as well as phytoextractions (Afewerky et al. 2021).

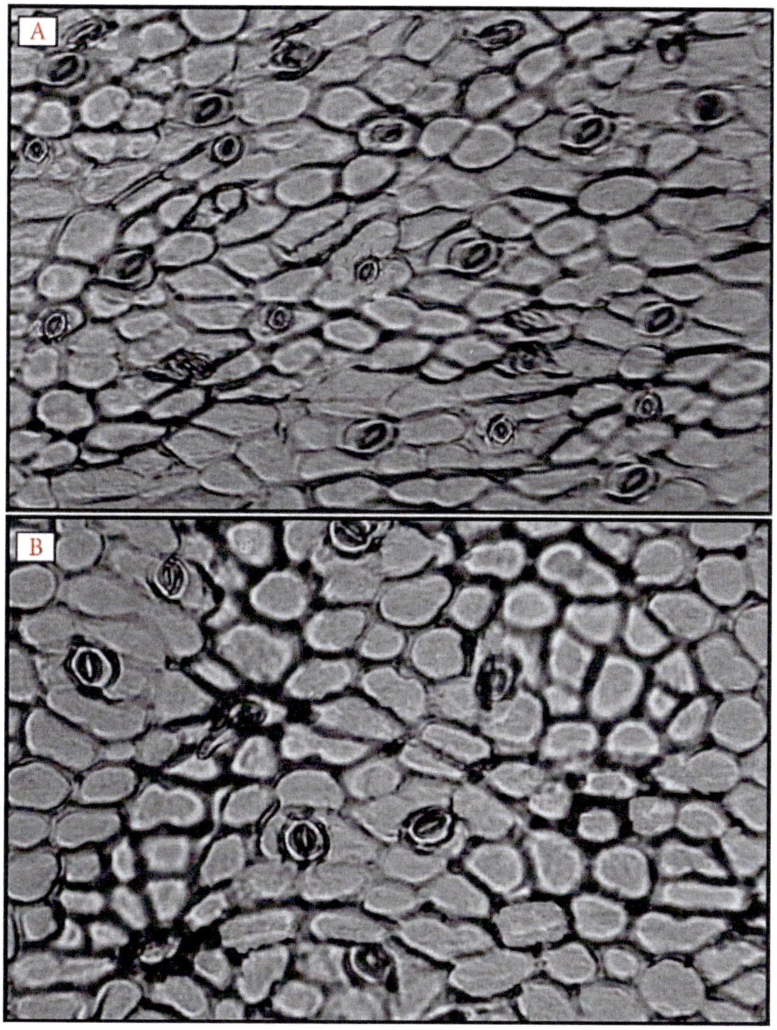

Fig. 2.50 (**a**) Abaxial surface; (**b**) Adaxial surface of *Withania somnifera* (L.) Dunal

2.40 *Ziziphus nummularia* (Burm.f.) Wight & Arn.

Family: Rhamnaceae

Synonym(s): *Rhamnus microphylla* Roxb.; *Rhamnus nummularia* Burm.f.; *Ziziphus microphylla* Roxb.; *Ziziphus nummularia* var. *glabrescens* Bhandari & Bhansali; *Ziziphus rotundifolia* Lam. (Malik et al. 2021).

English name: Wild Jujube (Malik et al. 2021).

Flowering period: July–September (Malik et al. 2021).

Distribution: Afghanistan, India, Iran, Iraq, Nepal, Pakistan, Palestine, and Saudi Arabia (Malik et al. 2021).

Altitude: Up to 1700 m (Malik et al. 2021).

Habit: Shrub

Ecological type: Xerophyte

Habitat: In dry deciduous forests and scrub jungles (Malik et al. 2021).

Phytochemicals: Flavonoids, alkaloids, glycosides, pectin, polysaccharides, peptide alkaloids, saponins, sterols, tannins, and fatty acids (Goyal et al. 2012, 2013; Hussain et al. 2017; Malik et al. 2021).

Potential biological activity: Antitumor, anthelmintic, antibacterial, analgesic, anti-inflammatory, anti-diarrheal, antisecretory, anti-spasmodic, and anti-ulcer effects (De Boer et al. 2005; Bachaya et al. 2009; Kumar et al. 2011; Goyal et al. 2013; Hussain et al. 2017; Malik et al. 2021).

Toxicity: Genotoxicity and cytotoxicity (Padalia and Chanda 2017; Malik et al. 2021).

Botanical description: Straggling shrub, 2–5 m high; branches divaricate, flexuous, tomentose when young; spines paired, one straight, very sharp, nearly as long as petiole, the other shorter, hooked; leaves ovate, elliptic or orbicular, oblique or rounded at base, serrate at the margin, acute or obtuse at apex, 1–2.5 × 0.5–1.8 cm, glabrous or tomentose above, tomentose to white-wooly beneath, 3-nerved at the base; inflorescences axillary, 10–20 flowered cymes, short-peduncled or not; flowers 4–5 mm across; calyx lobes deltoid, acute, keeled nearly to base, 1.7–2.2 mm long; petals obovate-spatulate, rounded, or truncate at apex, 1–1.5 mm long; stamens 0.8–1.2 mm long; disc faintly 10-lobed; ovary 2-loculed; style 2-cleft; drupes globose, ca 0.8 cm, woody, glabrous, shining, and blackish-red when ripe; seeds 1 or 2 compressed, black (Malik et al. 2021) (Fig. 2.51).

Fig. 2.51 *Ziziphus nummularia* (Burm.f.) Wight & Arn.

Pollen features: Polar length: 19.74–25.13 µm, equatorial length: 21.54–32.31 µm, exine thickness: 0.71–1.79 µm, tectum: strio-rugulate (Perveen and Qaiser 2005).

Anatomical features: According to the findings of Dinarvand and Zarinkamar (2006), both leaf surfaces are pubescent with simple short trichome; epidermis includes 1 layer of polygonal cells with thick outer wall and thick cuticle on both surfaces, stomata are sunken, visible in crypts, the papilla are visible on both surfaces more frequently on abaxial surface; mesophyll is isobilateral with 3–4 layers of long palisade cells adaxially and 2–3 layers of short palisade cells abaxially; the vascular bundles are collateral, surrounded by bundle sheaths with thin cell walls; the stem is spherical in shape in cross section, showing secondary development; also covered by thin walled unicellular trichomes; single epidermis includes polygonal cells with a thick outer wall and brown part; the cortex includes layers of sub epidermal collenchyma and sclerenchyma cells; in the vascular cylindrical, secondary development is important, it is surrounded by periphloematic fibers; the pith is small and includes parenchymatous cells in the center of stem (Dinarvand and Zarinkamar 2006).

Part used: Fruit, root (Malik et al. 2021).

Mode of utilization: Decoction, infusion (Malik et al. 2021).

Route of administration: Oral

Disease treated: Joint pain

Traditional uses: Leaves and fruits are used against the inflammation of gums, for indigestion, diarrhea, dysentery, cold and tonic (Kapoor and Arora 2014; Malik et al. 2021); whereas unripe fruits are prescribed in the burning sensations, management of vomiting, and as a tonic, the dried fruits are reported to be useful as sedative, stomachic, anodyne, refrigerant, as anticancer and in the treatment of chronic fatigue, bronchitis burns, diarrhea, hysteria, anemia, loss of appetite and pharyngitis (Hussain et al. 2017; Malik et al. 2021).

Recipes: (Malik et al. 2021)

- 30 mL of decoction of 50 g fruits used for 4–5 days everyday daily at night for joint pains.
- 40–50 g of roots dipped in 200 mL water for full one night, infusion made and used for joint pain for 14 days.

Potential economic significance: A multipurpose species valued for edible fruits, leaves as forage, branches for fencing, wood as fuel, for construction, and furniture (Pandey et al. 2010; Malik et al. 2021).

References

Abbasi AM, Shah MH, Khan MA (2015) Wild edible vegetables of lesser Himalayas. Springer International Publishing

Abd el Ghani MM (2000) Vegetation composition of Egyptian inland salt marshes. Bot Bull Acad Sinica 41:305–314

Abd El-Gawad AM, El-Amier YA (2017) Anatomical features of three perennial swampy plants of Poaceae, grown on the water stream banks in Nile Delta, Egypt. J Med Bot 1:58–64

Abdulrahaman AA, Oyedotun RA, Oladele FA (2011) Diagnostic significance of leaf epidermal features in the family Cucurbitaceae. Insight Bot 1(2):22–27

Adhikesavan C, Rajagopal K, Rajaduari JS (2015) Production and characterization of biodiesel from *Acacia nilotica* seeds. Int J ChemTech Res 8(2):854–859

Afewerky HK, Ayodeji AE, Tiamiyu BB et al (2021) Critical review of the *Withania somnifera* (L.) Dunal: ethnobotany, pharmacological efficacy, and commercialization significance in Africa. Bull Natl Res Cent 45(1):1–16

Agra M, Silva K, Basılio I et al (2008) Survey of medicinal plants used in the region Northeast of Brazil. Rev Bras Farmacogn 18:472–508

Ahatovic A, Čakar J, Subašić M et al (2020) *Plantago lanceolata* L. from serpentine soils in central bosnia tolerates high levels of heavy metals in soil. Water Air Soil Pollut 231:169

Ahmad A, Pillai KK, Najmi AK et al (2002) Evaluation of hepatoprotective potential of jigrine post-treatment against thioacetamide induced hepatic damage. J Ethnopharmacol 79:35–41

Ahmad F, Khan MA, Ahmad M et al (2011) Taxonomic application of foliar anatomy in grasses of tribe Eragrostideae (Poaceae) from salt range of Pakistan. Pak J Bot 43(5):2277–2284

Ahmad M, Sultana S, Fazli Hadi S et al (2014) An ethnobotanical study of medicinal plants in high mountainous region of Chail valley (District Swat-Pakistan). J Ethnobiol Ethnomed 10(1):1–18

Ahmad N, Shinwari ZK, Hussain J, Perveen R (2015) Phytochemicals, antibacterial and antioxidative investigations of *Alhagi maurorum* medik. Pak J Bot 47(1):121–124

Ahmad S, Zafar M, Ahmad M et al (2020) Seed morphology using SEM techniques for identification of useful grasses in Dera Ghazi Khan, Pakistan. Microsc Res Tech 83(3):249–258

Ahmed B (2007) Research on the production of natural vinegar from date palm sap. Available online at: www.rib-bangladesh.org/vinegar_research.php

Ahmed OH, Hamad MN, Jaafar NS (2017) Phytochemical investigation of *Chenopodium murale* (Family: Chenopodiaceae) cultivated in Iraq, isolation and identification of scopoletin and gallic acid. Asian J Pharm Clin Res 10(11):70–77

Akram M, Asif HM, Akhtar N et al (2011) *Tribulus terrestris* Linn.: a review article. J Med Plants Res 5(16):3601–3605

Al Hassan M, Gohari G, Boscaiu M et al (2015) Anatomical modifications in two Juncus species under salt stress conditions. Notulae Botanicae Horti Agrobotanici Cluj-Napoca 43(2):501–506

Al-Asmari AK, Al-Elaiwi AM, Athar MT et al (2014) A review of hepatoprotective plants used in Saudi traditional medicine. Evid Based Complement Alternat Med 2014:Article ID 890842

Aleem A, Janbaz KH (2017) Ethnopharmacological evaluation of *Cenchrus ciliaris* for multiple gastrointestinal disorders. Bangladesh J Pharmacol 12(2):125–132

Alhourani N, Kasabri V, Bustanji Y et al (2018) Potential antiproliferative activity and evaluation of essential oil composition of the aerial parts of *Tamarix aphylla* (L.) H. Karst.: a wild grown medicinal plant in Jordan. Evid Based Complement Alternat Med 2018:Article ID 9363868

Ali SI (1983) Asclepiadaceae. In: Nasir E, Ali SI (eds) Flora of Pakistan-Karachi, Vol, vol 150, pp 1–165

Ali M, Iqbal IM, Shabbir A et al (2020) Ethnomedicinal studies on aquatic plants of tehsil Shakargarh, Punjab, Pakistan. J Med Plants 8(1):15–19

Aliotta G, Greca MD, Monaco P et al (1990) In vitro algal growth inhibition by phytotoxins of *Typha latifolia* L. J Chem Ecol 16(9):2637–2646

Al-Khalil S (1995) A survey of plants used in Jordanian traditional medicine. Int J Pharmacogn 33(4):317–323

Al-Massarani S, Al-Enzi F, Al-Tamimi M et al (2016) Composition & biological activity of *Cyperus rotundus* L. tuber volatiles from Saudi Arabia. Nat Volatiles Essent Oils 3(2):26–34
Alothman EA, Awaad AS, Al-Qurayn NA et al (2018) Anticancer effect of *Cenchrus ciliaris* L. Saudi Pharm J 26(7):952–955
Al-Oudat M, Khatib Salkini A, Tiedeman J (2005) Major native plant species in the Knanasser area, Syria (Al-Hass and Shbeith mountains). ICARDA, Aleppo, p 147
Alshehri SA, Wahab S, Abullais SS et al (2022) Pharmacological efficacy of *Tamarix aphylla*: a comprehensive review. Plants 11(1):118
Al-Snafi AE (2016) The chemical constituents and therapeutic importance of *Cressa cretica*-A review. IOSR J Pharm 6(6):39–46
Altay V, Çelik O (2011) Antakya semt pazarlarındaki bazı doğal bitkilerin etnobotanik yönden araştırılması. Biyoloji Bilimleri Araştırma Dergisi 4(2):137–139
Altay V, Karahan F (2012) Tayfur Sökmen Kampüsü (Antakya-Hatay) ve çevresinde bulunan bitkiler üzerine etnobotanik bir araştırma. Karadeniz Fen Bilimleri Dergisi 3(2):13–28
Altay V, Ozturk M (2021) The genera *Salsola* and *Suaeda* (Amaranthaceae) and their value as fodder. In: Grigore M-N (ed) Handbook of halophytes. Springer Nature, Cham. https://doi.org/10.1007/978-3-030-17854-3_97-1
Altay V, Karahan F, Sarcan YB, Ilçim A (2015) An ethnobotanical research on wild plants sold in Kırıkhan district (Hatay/Turkey) herbalists and local markets. Biol Divers Conserv 8(2):81–91
Alwadie H (2002) Ultrastructure of the pollen grains of *Withania somnifera* (L.) Dunal (Solanaceae), A study from Saudi Arabia. Taeckholmia 22(1):115–119
Al-Watban AA, Al-Mogren E, Doaigey AR, El Zaidy M (2013) Pollen morphology of seven wild species of *Acacia* in Saudi Arabia. African J Plant Sci 7(12):602–607
Amamou F, Bouafia M, Chabane-Sari D et al (2011) *Citrullus colocynthis*: a desert plant native in Algeria, effects of fixed oil on blood homeostasis in Wistar rat. J Nat Prod Plant Resour 1:1–7
Andhiwal CK, Kishore K (1984) Sterols of *Salsola-foetida*. J Indian Chem Soc 61(8):729–730
Andrzejewska-Golec E (1992) Hair morphology in *Plantago* sect. Coronopus (Plantaginaceae). Plant Syst Evol 179:107–113
Angelini P, Bricchi E, Gigante D et al (2014) Pollen morphology of some species of Amaranthaceae s. lat. common in Italy. Flora Mediterr 24:247–272
Anis M, Sharma M, Iqbal M (2000) Herbal ethnomedicine of the Gwalior Forest Division in Madhya Pradesh, India. Pharm Biol 38:241–253
Ansari AA, Gill SS, Gill R, Lanza GR, Newman L (eds) (2018) Phytoremediation – management of environmental contaminants, vol 6. Springer Nature, Cham
Antonielli M, Pasqualini S, Batini P et al (2002) Physiological and anatomical characterization of *Phragmites australis* leaves. Aquat Bot 72(1):55–66
Aphale AA, Chibba A, Kumbhakarna NR et al (1998) Subacute toxicity study of the combination of ginseng (*Panax ginseng*) and ashwagandha (*Withania somnifera*) in rats: a safety assessment. Indian J Physiol Pharmacol 42(2):299–302
Arbelet-Bonnin D, Ben-Hamed-Louati I, Laurenti P et al (2019) Cakile maritima, a promising model for halophyte studies and a putative cash crop for saline agriculture. Adv Agron 155:45–78
Arı S, Temel M, Kargıoğlu M (2015) Ethnobotanical uses of *Phragmites australis* in Afyonkarahisar Province of Western Anatolia (Turkey). GJRA Glob J Res Anal 4(6):179–181
Arora M, Siddiqui AA, Paliwal S, Sood P (2014) A phyto-pharmacological overview on *Salvadora oleoides* Decne. Indian J Nat Prod Resour 5(3):209–214
Ashfaq S, Ahmad M, Zafar M et al (2020) Pollen morphology of family Solanaceae and its taxonomic significance. An Acad Bras Cienc 92(3):e20181221
Ashour OM, Abdel-Naim AB, Abdallah HM et al (2012) Evaluation of the potential cardioprotective activity of some saudi plants against doxorubicin toxicity. Zeitschrift für Naturforschung C 67:297–307
Ashraf MA, Mahmood K, Yusoff I, Qureshi AK (2013) Chemical constituents of *C. ciliaris* L. from the Cholistan desert, Pakistan. Arch Biol Sci 65:1473–1478

Ashraf MT, Fang C, Bochenski T et al (2016) Estimation of bioenergy potential for local biomass in the United Arab Emirates. Emir J Food Agric 28(2):99–106
Aynehchi Y, Salehi Sormaghi MH et al (1981) Survey of Iranian Plants for saponins, alkaloids, flavonoids and tannins. I. Q J Crude Drug Res 19(2–3):53–63
Babre NP, Gouda TS, Gowrishankar NL (2020) Hepatoprotective activity of against carbon tetrachloride induced hepatic damage *Pentatropis nivalis* in rats. Asian J Pharm Pharmacol 6(6):383–388
Bachaya HA, Iqbal Z, Khan MN, Jabbar A (2009) Anthelmintic activity of *Ziziphus nummularia* (bark) and *Acacia nilotica* (fruit) against *Trichostrongylid nematodes* of sheep. J Ethnopharmacol 123:325–329
Bahramsoltani R, Kalkhorani M, Zaidi SMA et al (2020) The genus *Tamarix*: traditional uses, phytochemistry, and pharmacology. J Ethnopharmacol 246:112245
Bangarwa KS (2008) Exploring Capparis decidua for livelihood and wasteland development. Asia-Pac Agrofor Newsl 32:3–5
Barman C, Singh VK, Tandon R (2018) Reproductive biology of *Salvadora oleoides* Decne. (Salvadoraceae). Int J Plant Reproduct Biol 10(1):69–76
Barness G, Rodriguez Zaragoza S, Shmueli I, Steinberger Y (2009) Vertical distribution of a soil microbial community as affected by plant ecophysiological adaptation in a desert system. Microb Ecol 57:36e49
Baum BR (1978) The genus *Tamarix*. Israel Academy of Sciences and Humanities, Jerusalem
Belcheff E (2012) A medical intuitive reveals the wonders of Purslane. Polished Publishing Group
Bennani-Kabchi N, El Bouayadi F, Kehel L et al (1999) Effect of *Suaeda fruticosa* aqueous extract in the hypercholesterolaemic and insulin-resistant sand rat. Therapie 54(6):725–730
Bennett AE, Bever JD, Bowers MD (2009) Arbuscular mycorrhizal fungal species suppress inducible plant responses and alter defensive strategies following herbivory. Oecologia 160:771–779
Bhandari MM (1978) Flora of the Indian desert. Scientific Publishers, Jodhpur, p 51
Bianco VV, Santamaria P, Elia A et al (1998) Nutritional value and nitrate content in edible wild species used in southern Italy. Acta Hortic 467:71–87
Bibi H, Afzal M, Muhammad A et al (2014) Morphological and anatomical studies on selected Dicot xerophytes of District Karak, Pakistan. Am Eurasian J Agric Environ Sci 14:1201–1212
Bibi H, Afzal M, Kamal M et al (2015) Morphological and anatomical characteristics of selected dicot xerophytes of district Karak, Khyber Pakhtunkhwa, Pakistan. Middle-East J Sci Res 23(4):545–557
Biradar NV, Mahabale TS (1969) Studies on palms: fruits, seeds and seed germination in the genus Phoenix L. Proc Indian Acad Sci Chem Sci B 70(2):55–65
Bisrat SA, Mullen BF, Grigg AH, Shelton HM (2004) Net primary productivity and rainfall use efficiency of pastures on reconstructed land following open-cut coal mining in central Queensland, Australia. Trop Grassl 38(1):47–55
Blumenthal-Goldschmidt S, Poljakoff-Mayber A (1968) Effect of substrate salinity on growth and on submicroscopic structure of leaf cells of *Atriplex halimus* L. Aust J Bot 16:469e478
Boakye YD, Shaheen S, Nawaz H et al (2017) Artemisia scoparia: a review on traditional uses, phytochemistry and pharmacological properties. Int J Chem Biochem Sci 12:92–97
Bonham AJ (1980) Bank protection using emergent plants against boat wash in rivers and canals. Report, Hydraulics Research Station, UK
Bose U, Bala V, Ghosh TN et al (2011) Antinociceptive, cytotoxic and antibacterial activities of *Cleome viscosa* leaves. Revista Brasileira de Farmacognosia 21:165–169
Bourhia M, Messaoudi M, Bakrim H et al (2020) *Citrullus colocynthis* (L.) Schrad: chemical characterization, scavenging and cytotoxic activities. Open Chem 18(1):986–994
Bouzid A, Benabdeli K (2011) Contribution to the assessment of green biomass of *Atriplex halimus* plantation in arid western Algeria (region of Naaama). Rev D Ecol-La Tierre et la Vie 66:303e308
Brajeshwar (2001–2002) Gokshura. Wastelands News 17(2):46

Brandis D (1972) The forest-flora of north-west and central India. Bishen Singh Mahendra Pal Singh, Dehradun

Bughio SH, Samejo MQ, Memon S et al (2017) Chemical composition of the essential oils from *Tamarix dioica* and determination of its antibacterial activity. Int J Food Prop 20(Suppl 3):S2660–S2667

Burrows GE, Shaik RS (2015) Comparative developmental anatomy of the taproot of the Cucurbitaceous vines *Citrullus colocynthis* (perennial), *Citrullus lanatus* (annual) and *Cucumis myriocarpus* (annual). Aust J Bot 62(7):537–545

Caparrós PG, Ozturk M, Gul A et al (2022) Halophytes have potential as heavy metal phytoremediators: a comprehensive review. Environ Exp Bot 193:104666

Ceccanti C, Finimundy TC, Melgar B et al (2022) Sequential steps of the incorporation of bioactive plant extracts from wild Italian *Plantago coronopus* L. and *Cichorium intybus* L. leaves in fresh egg pasta. Food Chem 2022:132462

Çelik TA, Aslantürk ÖS (2006) Anti-mitotic and anti-genotoxic effects of *Plantago lanceolata* aqueous extract on *Allium cepa* root tip meristem cells. Biologia 61(6):693–697

Chakou FZ, Boual Z, Hadj MDOE et al (2021) Pharmacological ınvestigations in traditional utilization of *Alhagi maurorum* Medik. in Saharan Algeria: in vitro study of anti-ınflammatory and antihyperglycemic activities of water-soluble polysaccharides extracted from the seeds. Plants 10(12):2658

Chandel AK, Singh OV, Rao VL et al (2011) Bioconversion of novel substrate *Saccharum spontaneum*, a weedy material, into ethanol by Pichiasti-pitis NCIM3498. Bioresour Technol 102:1709–1714

Chang K-S, Shin E-H, Park C, Ahn Y-H (2012) Contact and fumigant toxicity of Cyperus rotundus steam distillate constituents and related compounds to insecticide-susceptible and-resistant Blattella germanica. J Med Entomol 49(3):631–639

Chatterjee S, Srivastava S, Khalid A et al (2010) Comprehensive metabolic fingerprinting of *Withania somnifera* leaf and root extracts. Phytochemistry 71:1085–1094

Chaturvedi M, Datta K (2001) Pollen morphology in *Saccharum* L. (Poaceae)-wild and cultivated sugar cane species. Feddes Repert 112(5–6):387–390

Chaudhary S, Kumar R (2015) Ethnomedicinal plants of the district Bijnor (U.P.) India. J Indian Bot Soc 94:97–103

Chaudhary G, Singh LK, Ghosh S (2012) Alkaline pretreatment methods followed by acid hydrolysis of *Saccharum spontaneum* for bioethanol production. Bioresour Technol 124:111–118

Chehregani A, Noori M, Yazdi HL (2009) Phytoremediation of heavy-metal-polluted soils: screening for new accumulator plants in Angouran mine (Iran) and evaluation of removal ability. Ecotoxicol Environ Saf 72(5):1349–1353

Chen CJ, Wang WY, Wang XL et al (2009) Anti-hypoxic activity of the ethanol extract from *Portulaca oleracea* in mice. J Ethnopharmacol 124(2):246–250

Chen T, Wang J, Li Y et al (2010) Sulfated modification and cytotoxicity of *Portulaca oleracea* L. polysaccharides. Glycoconj J 27(6):635–642

Chen HY, Lin YH, Su IH et al (2014) Investigation on Chinese herbal medicine for primary dysmenorrhea: implication from a nationwide prescription database in Taiwan. Complement Ther Med 22(1):116–125

Chhatre S, Nesari T, Somani G et al (2014) Phytopharmacological overview of *Tribulus terrestris*. Pharmacogn Rev 8(15):45

Chikhi I, Allali H, Dib MEA et al (2014) Antidiabetic activity of aqueous leaf extract of *Atriplex halimus* L.(Chenopodiaceae) in streptozotocin-induced diabetic rats. Asian Pac J Trop Dis 4(3):181–184

Chopra RN, Chopra IC, Handa KL, Kapur LD (1958) Indigenous drugs of India. U. N. Dhur & Sons Private Limited, Calcutta

Choudhary SP, Sharma DK (2014) Bioactive constituents, phytochemical and pharmacological properties of *Chenopodium album*: a miracle weed. Int J Pharmacogn 1:545–552

Cornara L, Ambu G, Alberto A et al (2022) Characterization of ingredients incorporated in the traditional mixed-salad of the Capuchin Monks. Plants 11(3):301

Dafni A, Lev E (2002) The doctrine of Signatures in present-day Israel. Econ Bot 56:328–334

Das U, Layek U (2021) Pharmacognostic and phytochemical evaluation of *Cleome viscosa* L. (Cleomaceae). Int J Creat Res Thoughts 9(9):491–499

Davis PH (1965) Flora of Turkey and the East Aegean islands, vol 1. University Press, Edinburgh

Davis PH (1985) Flora of Turkey and the East Aegean islands, vol 9. University Press, Edinburgh

Davy AJ, Scott R, Cordazzo CV (2006) Biological flora of the British Isles: *Cakile maritima* Scop. J Ecol 94:695–711

de Blas IO (1992) Morfología de *Limonium sinuatum* (L.) Miller (Plumbaginaceae). Rev Biol Trop 40(1):11–17

De Boer HJ, Kool A, Broberg A et al (2005) Anti-fungal and anti-bacterial activity of some herbal remedies from Tanzania. J Ethnopharmacol 96:461–469

de Brito Damasceno GA, Ferrari M, Giordani RB (2017) *Prosopis juliflora* (Sw) DC., an invasive specie at the Brazilian Caatinga: phytochemical, pharmacological, toxicological and technological overview. Phytochem Rev 16(2):309–331

de Souza Mesquita LM, Colpo KD, da Rocha CQ et al (2017) Anatomical differentiation and metabolomic profiling: a tool in the diagnostic characterization of some medicinal *Plantago* species. Rev Bras Bot 40(3):801–810

Dehghani M, Djamali M, Akhani H (2021) Pollen morphology of the subfamily Salicornioideae (Chenopodiaceae) in Eurasia and North Africa. Palynology 45(2):245–258

Della A, Paraskeva-Hadjichambi D, Hadjichambis AC (2006) An ethnobotanical survey of wild edible plants of Paphos and Larnaca countryside of Cyprus. J Ethnobiol Ethnomed 2(1):1–9

Dely-Draskovits A, Vasarhely T, Bachli G (1992) The importance of reed beds from the entomological point of view. Mitt Entomol Ges Basel 42(2):46–52

Demirezen D, Aksoy A (2004) Accumulation of heavy metals in *Typha angustifolia* (L.) and *Potamogeton pectinatus* (L.) living in sultan marsh (Kayseri, Turkey). Chemosphere 56(7):685–696

Deng X, Xia Y, Hu W et al (2010) Cadmium-induced oxidative damage and protective effects of n-acetyl-l-cysteine against cadmium toxicity in *Solanum nigrum* L. J Hazard Mater 180(1–3):722–729

Dhanani T, Shah S, Gajbhiye N, Kumar S (2017) Effect of extraction methods on yield, phytochemical constituents and antioxidant activity of *Withania somnifera*. Arab J Chem 10(3):S1193–S1199

Dinarvand M, Zarinkamar F (2006) Anatomy-taxonomy of the genus *Ziziphus* in Iran. Iran J Bot 12(1):36–41

Dobberteen RA, Nickerson NH (1991) Use of created cattail (*Typha*) wetlands in mitigation strategies. Environ Manag 15(4):797–808

Doležal J, Dvorský M, Börner A et al (2018) Anatomy, age and ecology of high mountain plants in Ladakh, the Western Himalaya. Springer Nature. https://doi.org/10.1007/978-3-319-78699-5

Dvorak J (1996) An example of relationships between macrophytes, macroinvertebrates and their food resources in a shallow eutrophic lake. Hydrobiologia 339(1–3):27–36

El Eraky WI (2001) Pharmacological effects of *Noaea mucronata*. Egypt J Med Lab Sci 10(1):13–23

El-Aasr M, Kabbash A, El-Seoud KAA et al (2016) Antimicrobial and immunomodulatory activities of flavonol glycosides isolated from *Atriplex halimus* L. herb. J Pharm Sci Res 8:1159–1168

El-Amier YA (2015) Morphological studies of the pollen grains for some hydrophytes in coastal Mediterranean lakes, Egypt. Egypt J Basic Appl Sci 2(2):132–138

El-Ghonemy A (1993) Encyclopedia of medicinal plants of the United Arab Emirates, vol 568. University of United Arab Emirates Press

Elkordy A, Faried A (2017) Pollen morphology and numerical analysis of *Tamarix* L.(Tamaricaceae) in Egypt and its systematic implication. Bangladesh J Plant Taxon 24(1):91–105

El-Rokiek K, El-Din S, Sharara F (2010) Allelopathic behaviour of *Cyperus rotundus* L. on both *Chorchorus olitorius* (broad leaved weed) and *Echinochloa crus-galli* (Grassy weed) associated with soybean. J Plant Prot Res 50(3):274–279
El-Sayed NH, Ishak MS, Kandil FI, Mabry TJ (1993) Flavonoids of *Alhagi graecourm*. Pharmazie 48:68–69
EMPPO (European and Mediterranean Plant Protection Organization) (2019) Prosopis juliflora (Sw.) DC. Bull OEPP/EPPO Bull 49(2):290–297
Fahmy IR (1963) Drug plants of Egypt. Planta Med 11:202–224
FAO (2002) Afghanistan. Country pasture/forage resources profiles. AGPG Crop and Grassland Services. http://www.fao.org/ag/AGPG/doc/pasture/pature.htm
Fontana A, Reichelt M, Hempel et al (2009) The effects of arbuscular mycorrhizal fungi on direct and indirect defense metabolites of *Plantago lanceolata* L. J Chem Ecol 35:833–843
Fuochi V, Barbagallo I, Distefano A et al (2019) Biological properties of *Cakile maritima* Scop. (Brassicaceae) extracts. Eur Rev Med Pharmacol Sci 23(5):2280–2292
Galati E, Monforte M, Forestieri A et al (1999) *Salvadora persica* L.: hypolipidemic activity on experimental hypercholesterolemia in rat. Phytomedicine 6(3):181–185
Galvez M, Martın-Cordero C, Lopez-Lazaro M et al (2003) Cytotoxic effect of Plantago spp. on cancer cell lines. J Ethnopharmacol 88(2–3):125–130
Garg A, Mittal SK, Kumar M et al (2014) Phyto-pharmacological study of *Salvadora oleoides*. Int J Bioassays 3:1714–1717
Garti N, Slavin Y, Aserin A (1999) Surface and emulsification properties of a new gum extracted from *Portulaca oleracea* L. Food Hydrocoll 13(2):145–155
Gescher K, Deters AM (2011) *Typha latifolia* L. fruit polysaccharides induce the differentiation and stimulate the proliferation of human keratinocytes in vitro. J Ethnopharmacol 137(1):352–358
Ghahraman A, Nourbakhsh N, Mehdi GK, Atar F (2007) Pollen morphology of *Artemisia* L.(Asteraceae) in Iran. Iran J Bot 13(1):21–29
Ghangro IH, Ghangro AB, Channa MJ (2015) Nutritional assessment of non-conventional vegetable *C. decidua* flower. Rawal Med J 40(2):214–216
Ghasemi Pirbalouti A, Momeni M, Bahmani M (2013) Ethnobotanical study of medicinal plants used by kurd tribe in Dehloran and Abdanan districts, Ilam province, Iran. Afr J Tradit Complement Altern Med 10:368–385
Global (2014) Sustainability assessment system. An overview. Gulf Organization for Research and Development, Doha, pp 1–35
Gonçalves S, Romano A (2016) The medicinal potential of plants from the genus *Plantago* (Plantaginaceae). Ind Crop Prod 83:213–226
Goncharov MY, Yakovlev GP, Vitovskaya GA (2001) Composition of polysaccharides from aboveground part of *Alhagi maurorum* Medic. Rastitel'nye Resursy 37:60–63
Goyal M, Nagori BP, Sasmal D (2012) Review on ethnomedicinal uses, pharmacological activity and phytochemical constituents of *Ziziphus mauritiana (Z. jujuba* Lam., non mill). Spatula DD 2(2):107–116
Goyal M, Ghosh M, Nagori BP, Sasmal D (2013) Analgesic and anti-inflammatory studies of cyclopeptide alkaloid fraction of leaves of *Ziziyphus nummularia*. Saudi J Biol Sci 20:365–371
Grace JB, Harrison JS (1986) The biology of Canadian weeds. 73. *Typha latifolia* L., *Typha angustifolia* L. and *Typha x glauca* Godr. Can J Plant Sci 66(2):361–379
Graveson R (2012) The Plants of Saint Lucia (in the Lesser Antilles of the Caribbean). http://www.saintlucianplants.com
Gray KR, Biddlestone AJ (1995) Engineered reed-bed systems for wastewater treatment. Trends Biotechnol 13(7):248–252
Grieve CM, Poss JA, Grattan SR et al (2005) Productivity and mineral nutrition of *Limonium* species irrigated with saline wastewaters. HortScience 40(3):654–658
Griffin GF, Smith DMS, Morton SR et al (1989) Status and implications of the invasion of tamarisk (*Tamarix aphylla*) on the Finke River, Northern Territory, Australia. J Environ Manag 29(4):297–315

Guil JL, Torija ME, Gimenez JJ et al (1996) Oxalic acid and calcium determination in wild edible plants. J Agric Food Chem 44:1821–1823
Guil JL, Rodríguez-Garcí I, Torija E (1997) Nutritional and toxic factors in selected wild edible plants. Plant Foods Hum Nutr 51(2):99–107
Gul B, Ansari R, Ali H et al (2014) The sustainable utilization of saline resources for livestock feed production in arid and semi-arid regions: a model from Pakistan. Emirates J Food Agric 26(12):1032–1045
Gull T, Sultana B, Bhatti IA et al (2015) Antibacterial potential of *Capparis spinosa* and *Capparis decidua* extracts. Int J Agric Biol 17:727–733
Gurudeeban S, Ramanathan T, Satyavani K (2011) Characterization of volatile compounds from bitter apple (*Citrullus colocynthis*) using GC-MS. Int J Chem Analyt Sci 2(8):108–110
Hamdi SM, Assadi M, Segarra-Moragues JG (2010) Pollen morphology of Iranian species of *Typha* L.(Typhaceae) and its taxonomic significance. Feddes Repert 121(1–2):85–96
Hameed A, Hussain T, Gulzar S et al (2012) Salt tolerance of a cash crop halophyte *Suaeda fruticosa*: biochemical responses to salt and exogenous chemical treatments. Acta Physiol Plant 34(6):2331–2340
Hand R (2003) Supplementary notes to the flora of Cyprus III. BGBM Berlin-Dahlem (2ed.). Willdenowia 33:305–325
Hanif Z, Naeem M, Ali HH et al (2017) Effect of environmental factors on germination of *Salsola foetida*: potential species for rehabilitation of degraded rangelands. Rangel Ecol Manag 70(5):638–643
Hashem A, Abd-Allah EF, Alqarawi AA et al (2015) Impact of plant growth promoting *Bacillus subtilis* on growth and physiological parameters of *Bassia indica* (Indian Bassia) grown under salt stress. Pak J Bot 47(5):1735–1741
Hashem A, Abd-Allah EF, Alqarawi AA et al (2019) Role of calcium in AMF-mediated alleviation of the adverse impacts of cadmium stress in *Bassia indica* [Wight] A.J. Scott. Saudi J Biol Sci 26(4):828–838
Hatfield AW (1971) How to enjoy your weeds. Sterling Publishing Co., Inc., New York
Hayrapetyan A, Sonyan H (2021) Palynomorphological peculiarities of some representatives of the subfamily Salsoloideae (Chenopodiaceae) in South Transcaucasia. I. *Seidlitzia* Bunge and *Noaea* Moq. Biol J Armenia 2(73):42–51
He X, Mouratov S, Steinbeger Y (2002) Spatial distribution and colonization of arbscular mycorrhizal fungi under the canopies of desert halophytes. Arid Land Res Manag 16:149–160
Hebbar SS, Harsha VH, Shripathi V et al (2004) Ethnomedicine of Dharwad district in Karnataka, India-plants used in oral health care. J Ethnopharmacol 94:261–266
Heneidak S, Hassan AR (2005) Taxonomic significance of the seed characters of certain species of tribe Asclepiadeae in Egypt. Taeckholmia 25(1):91–109
Heneidy SZ, Halmy MWA, Bidak LM (2017) The ethnobotanical importance and conservation value of native plants in eastern Arabian Peninsula. Feddes Repert 128(3–4):105–128
Henni M, Mehdadi Z (2012) Preliminary assessment of soil and floristic characteristics of degraded steppes of white armoise rehabilitated by the planting of *Atriplex* in the region of Saida (western Algeria). Acta Bot Gall 159:43–52
Holm LG, Plucknett DL, Pancho JV, Herberger JP (1977) The world's worst weeds. Distribution and biology. University Press of Hawaii, Honolulu
Holm LG, Pancho JV, Herberger JP, Plucknett DL (1991) A geographic atlas of world weeds. Krieger Publishing Company, Malabar
Holm L, Doll J, Holm E et al (1997) World weeds. Natural histories and distribution. Wiley, New York
Hughes JB, Sousa JS, Barreto RA et al (2006) Cytotoxic effects of an extract containing alkaloids obtained from Prosopis juliflora Sw. D.C. (Algaroba) pods on glioblastoma cells. Rev Bras Sau'de Prod Anim 43:50–58
Hussain A (2020) The genus *Artemisia* (Asteraceae): a review on its ethnomedicinal prominence and taxonomy with emphasis on foliar anatomy, morphology, and molecular phylogeny. Proc Pakistan Acad Sci B Life Environ Sci 57(1):1–28

Hussain F, Ahmad B, Ilahi I (2010) Allelopathic effects of *Cenchrus ciliaris* L. and *Bothriochloa pertusa* (L.) A. Camus. Pak J Bot 42(5):3587–3604

Hussain AI, Rathore HA, Sattar MZ et al (2014) *Citrullus colocynthis* (L.) Schrad (bitter apple fruit): a review of its phytochemistry, pharmacology, traditional uses and nutritional potential. J Ethnopharmacol 155(1):54–66

Hussain SM, Khan A, Khan AU et al (2017) Pharmacological basis for medicinal use of *Ziziphyus nummularia* (Rhamnaceae) leaves in gastrointestinal disorders. Trop J Pharm Res 16(10):2379–2385

Ianovici N (2010) Histoanatomical studies on some halophytes from Romania-*Plantago coronopus*. In: Biomonitoring in urban environment view project cost action fa 1203-sustainable management of *Ambrosia artemisiifolia* in Europe (SMARTER); 2013–2017 view project. West University of Timişoara, Timisşoara

Idm'hand E, Msanda F, Cherifi K (2020) Ethnopharmacological review of medicinal plants used to manage diabetes in Morocco. Clin Phytosci 6(1):1–32

Imtiaz SM, Aleem A, Saqib F et al (2019) The potential involvement of an ATP-Dependent potassium channel-opening mechanism in the smooth muscle relaxant properties of *Tamarix dioica* Roxb. Biomol Ther 9(11):722

Inam-ur R, Daniel M, Henri R, Urs W (2011) Indigenous fodder trees can increase grazing accessibility for landless and mobile pastoralists in northern Pakistan. Pastor Res Policy Pract 1:2

Iqbal M, Ghouse AKM (1987) Anatomy of the vascular cambium of *Acacia nilotica* (L.) Del. var. *telia* Troup (Mimosaceae) in relation to age and season. Bot J Linn Soc 94(3):385–397

Iqbal J, Shah A, Sarvat R et al (2019) Documentation of folk herbal uses of medicinally important wild vegetables used by the tribal communities of Sargodha Region, Pakistan. Planta Daninha 37:e019189207

Iqbal U, Hameed M, Ahmad F (2021) Water conservation strategies through anatomical traits in the endangered arid zone species *Salvadora oleoides* Decne. Turk J Bot 45(2):140–157

Iranshahy M, Javadi B, Iranshahi M et al (2017) A review of traditional uses, phytochemistry and pharmacology of *Portulaca oleracea* L. J Ethnopharmacol 205:158–172

Ismail S, Nisar MF (2010) Ethnomedicinal survey for important plants of district Lodhran, Punjab, Pakistan. BIOL E-J Life Sci 1(3):52–58

Jaafar NS, Jaafar IS, Noori ZS (2021) *Cressa cretica* – pharmacognosy, and pharmacology (a review). Iraqi J Pharmaceut Sci 30(2):31–40

Jain RK (1994) Fuel wood characteristics of medium tree and shrub species of India. Bioresour Technol 47(1):81–84

Jain P, Jain S, Sharma S, Paliwal S (2018) Diverse application of *Phoenix sylvestris*: a potential herb. Agric Nat Resour 52(2):107–114

James AD (2011) *Alhagi maurorum* (FABACEAE). Dr. Duke's phytochemical and ethnobotanical databases. Retrieved 13 Nov 2011

Janbaz KH, Aslam N, Imran I, Jabeen Q (2021) Evaluation of anti-inflammatory, analgesic and antipyretic activities of *Salsola imbricata* Forssk in rats. J Anim Plant Sci 31(3):862–867

Janković T, Zdunić G, Beara I et al (2012) Comparative study of some polyphenols in Plantago species. Biochem Syst Ecol 42:69–74

Jansen PCM (2004) *Chenopodium album* L. – Vegetables. Plant Resour Trop Afr 2:178–180

Javed F, Jabeen Q (2021) *Salsola imbricata* Forssk. ameliorates acetic acid-induced inflammatory bowel disease by modulating dysregulated antioxidant enzyme system and cytokine signaling pathways in mice. Asian Pac J Trop Biomed 11(12):527

Javed S, Javaid A, Qureshi MZ (2018) Antifungal phytocomponents in n-butanol fraction of leaf extract of *Kochia indica* Wight. Int J Biol Biotechnol 15(4):661–666

Jianu LD, Bercu R, Popoviciu RD (2014) Anatomical features of the endengered plant *Cakile maritima* Scop. subsp. *euxina* (Pobed.) Nyar. Annales of West University of Timisoara. Series Biol 17(2):79

Joshi AR, Joshi K (2000) Indigenous knowledge and uses of medicinal plants by local communities of the Kali Gandaki Watershed Area, Nepal. J Ethnopharmacol 73:175–183

Kapoor LD (1990) Handbook of ayurvedic medicinal plants. CRC Press, Boca Raton

Kapoor BBS, Arora V (2014) Ethnomedicinal plants of Jaisalmer District of Rajasthan used in herbal and folk remedies. Int J Ethnobiol Ethnomed 1:1

Kasem WT (2016) Anatomical, pollen grains and seed exomorphic studies on five species of *Cleome* L. (Cleomaceae Bercht. & Presl) collected from South West of Saudi Arabia. J Plant Sci 4(2):29

Katewa SS, Galav PK (2006) Additions to the traditional folk herbal medicines from Shekhawati region of Rajasthan. Indian J Tradit Knowl 5(4):494–500

Katewa SS, Chaudhary BL, Jain A (2004) Folk herbal medicines from tribal area of Rajasthan, India. J Ethnopharmacol 92(1):41–46

Kayani S, Ahmad M, Zafar M et al (2014) Ethnobotanical uses of medicinal plants for respiratory disorders among the inhabitants of Gallies-Abbottabad, Northern Pakistan. J Ethnopharmacol 156:47–60

Khalid M, Siddiqui HH, Freed S (2011) Free radical scavenging and total phenolic content of *Saccharum spontaneum* L. root extracts. Int J Res Pharm Chem 1:1160–1166

Khan AU (1994) Appraisal of ethno-ecological incentives to promote conservation of *Salvadora oleoides* Decne: the case for creating a resource area. Biol Conserv 75(2):187–190

Khan KM, Maharvi GM, Abbaskhan A et al (2003) Three tyrosinase inhibitors and antioxidant compounds from *Salsola foetida*. Helv Chim Acta 86(2):457–464

Khan MA, Shinwari MI, Niazi HA (2005) Medicinal plants of Potohar region of Pakistan. Resource bases of medicinal plants in Potohar. http://www.telmedpak.com/agricultures.asp?a=medplantpak&b=med_plant8

Khan B, Ahmad SF, Bani S et al (2006) Augmentation and proliferation of T lymphocytes and Th-1 cytokines by *Withania somnifera* in stressed mice. Int Immunopharmacol 6:1394–1403

Khan AL, Hussain J, Hamayun M et al (2010) Secondary metabolites from *Inula britannica* L. and their biological activities. Molecules 15(3):1562–1577

Khan AW, Jan S, Parveen S et al (2012) Phytochemical analysis and enzyme inhibition assay of *Aerva javanica* for ulcer. Chem Cent J 6(1):1–6

Khan N, Ahmed M, Khan RA, Gul S (2019) Antioxidant, cytotoxicity activities and phytochemical analysis of *Chenopodium murale* (Linn.). Int J Bot 4:25–28

Khan F, Muhammad Z, Khan K et al (2020) Palynological investigation of allergenic and invasive weeds plants for biodiversity in District Lakki Marwat using scanning electron microscopy. Pakistan J Weed Sci Res 26(3):349–365

Khare C (2007) Indian medicinal plants. Springer, New York. https://doi.org/10.1007/978-0-387-70638-2_861

Khatun S, Ali MB, Hahn E-J, Paek K-Y (2008) Copper toxicity in *Withania somnifera*: growth and antioxidant enzymes responses of in vitro grown plants. Environ Exp Bot 64(3):279–285

Kirtikar KR, Basu BD (1987) Indian medicinal plants, 2nd edn, Dehradun

Kirtikar KR, Basu BD, Blatter E et al (1993) Indian medicinal plants, vol IV. Lalit Mohan Basu, Allahabad

Klimko M, Idzikowska K, Truchan M, Kreft A (2004) Pollen morphology of *Plantago* species native to Poland and their taxonomic implications. Acta Soc Bot Pol 73(4):315–325

Köbbing JF, Thevs N, Zerbe S (2013) The utilization of reed (*Phragmites australis*): a review. Mires Peat 13:1–14

Kokanova-Nedialkova Z, Nedialkov P, Nikolov S (2009) The genus *Chenopodium*: phytochemistry, ethnopharmacology and pharmacology. Pharmacogn Rev 3(6):280

Krisdianto K, Damayanti R (2007) Anatomical properties and fiber dimension of prickly Acacia (*Acacia nilotica* L.) from Baluran National Park. Indones J For Res 4(2):93–103

Ksouri R, Ksouri WM, Jallali I et al (2012) Medicinal halophytes: potent source of health promoting biomolecules with medical, nutraceutical and food applications. Crit Rev Biotechnol 32(4):289–326

Kulkarni AR, Mulani RM (2004) Indigenous palms of India. Curr Sci 86(12):1598–1603

Kumar R, Bhagat N (2012) Ethnomedicinal plants of district Kathua (J&K). Int J Med Arom Plants 2:603–611

Kumar S, Kumar V (2011) Evaluation of *Withania somnifera* L. (Dunal). (Solanaceae) leaf and root extracts as an antimicrobial agent-highly medicinal plants in India. Asian J Exp Biol Sci 2:155–157

Kumar CAS, Varadharajan R, Muthumani P et al (2010) Psychopharmacological studies on the stem of *Saccharum spontaneum*. Int J PharmTech Res 2(1):319–321

Kumar S, Garg VK, Kumar N et al (2011) Pharmacognostical studies on the leaves of *Ziziphus nummularia* (Burm. F.). Eur J Exp Biol 1:77–83

Kundu S (2009) A synopsis of Tamaricaceae in the Indian subcontinent: its distribution and endemism. Acta Bot Hungar 51(3–4):301–314

La Rosa A, Cornara L, Saitta A et al (2021) Ethnobotany of the Aegadian Islands: safeguarding biocultural refugia in the Mediterranean. J Ethnobiol Ethnomed 17(1):1–19

Landge SN, Shinde RD, Mistry MK (2021) On the correct identity and distribution of *Dactyloctenium scindicum* (Poaceae: Chloridoideae: Cynodonteae: Dactylocteniinae) in the Indo-Gangetic plains and Peninsular India and notes on other species in India. RHEEDEA J Indian Assoc Angiosp Taxon 31(4):282–295

Langar PN, Bakshi MPS (1990) Nutritive evaluation of fodders of Kashmir valley. Indian J Anim Sci 60(4):498–500

Le Dang Q, Lee GY, Choi YH et al (2010) Insecticidal activities of crude extracts and phospholipids from *Chenopodium ficifolium* against melon and cotton aphid, *Aphis gossypii*. Crop Prot 29(10):1124–1129

Lev E, Amar Z (2002) Ethnopharmacological survey of traditional drugs sold in the Kingdom of Jordan. J Ethnopharmacol 82(2–3):131–145

Li SY, Cao YS (1981) A preliminary study on the biological characters of Japanese hop (*Humulus scandens*) and its control. Acta Agric Univ Pek 7(3):45–46

Liguori I, Russo G, Curcio F et al (2018) Oxidative stress, aging, and diseases. Clin Interv Aging 13:757–772

Londonkar RL, Kattegouga UM, Shivsharanappa K, Hanchinalmath JV (2013) Phytochemical screening and in vitro antimicrobial activity of *Typha angustifolia* Linn leaves extract against pathogenic gram negative microorganisms. J Pharm Res 6(2):280–283

Madhavi HR, Neha KS, Saininand SA et al (2014) A holistic approach on review of *Solanum virginianum* L. RRJPPS 3(3):1–4

Mahdi MF, Abaas IS (2019) Cytotoxic activity of Iraqi *Cressa cretica*. Al-AJPS 19(1):95–102

Maheshwari JK (1963) The flora of Delhi. CSIR, New Delhi

Mahla HR, Rathore VS, Singh D, Singh JP (2013) *Capparis decidua* (Forsk.) Edgew.: an underutilized multipurpose shrub of hot arid region-distribution, diversity and utilization. Genet Resour Crop Evol 60(1):385–394

Mahmood A, Mahmood A, Hussain I, Kiyani WK (2011) Indigenous medicinal knowledge of medicinal plants of Barnala area, District Bhimber, Pakistan. Intl J Med Arom Plants 1:294–301

Malhotra SP, Dutta BK, Gupta R, Gaur YD (1966) Medicinal plants of the Indian arid zone. J Agric Trop Bot Appl 13(6):247–288

Mali RG (2010) *Cleome viscose* (wild mustard): a review on ethnobotany, phytochemistry, and pharmacology. Pharm Biol 48:105–112

Malik K, Ahmad M, Öztürk M et al (2021) Herbals of Asia – prevalent diseases and their treatments. Springer Nature, Cham. https://doi.org/10.1007/978-3-030-85222-1

Marak H, Biere A, Van Damme J (2000) Direct and correlated responses to selection on iridoid glycosides in *Plantago lanceolata* L. J Evol Biol 13(6):985–996

Marshall NA, Friedel M, van Klinken RD, Grice AC (2011) Considering the social dimension of invasive species: the case of buffel grass. Environ Sci Pol 14(3):327–338

Martin RM, Cox JR, Alston DG, Ibarra FF (1995) Spittlebug (Homoptera: Cercopidae) life cycle on buffelgrass in Northwestern Mexico. Ann Entomol Soc Am 88(4):471–478

McManus HA, Seago JL Jr, Marsh LC (2002) Epifluorescent and histochemical aspects of shoot anatomy of *Typha latifolia* L., *Typha angustifolia* L. and *Typha glauca* Godr. Ann Bot 90(4):489–493

Messina CM, Renda G, Laudicella VA et al (2019) From ecology to biotechnology, the study of the defense strategies of algae and halophytes (from Trapani Saltworks, NW Sicily) with a focus on antioxidants and antimicrobial properties. Int J Mol Sci 20(4):881

Miara MD, Souidi Z, Benhanifa K et al (2021) Diversity, natural habitats, ethnobotany and conservation of the flora of the Macta marches (North-West Algeria). Int J Environ Stud 78(5):817–837

Mirjalili MH, Mayano E, Mercedes B, Cusido RM, Palazón J (2009) Steroidal lactones from *Withania somnifera*, an ancient plant for novel medicine. Molecules 14:2373–2393

Mohammed MS, Khalid HS, Osman WJA, Muddathir AK (2014) A review on phytochemical profle and biological activites of three anti-infammatory plants used in sudanese folkloric medicine. Am J Pharm Tech Res 4(4):1–14

Mouhoub F, Ouafi S, Chabane D (2018a) Acute and subacute oral toxicity assessment of gender of Ahaggar's *Aerva javanica* in animal models. Int J Pharmacol 14(5):640–651

Mouhoub F, Chabane D, Ouafi S, Bouguedoura N (2018b) Impact of arid climate on sex distribution, morphological and anatomical organization of a medicinal herb *Aerva javanica* (Burm. F.) Juss. ex Schult (Amaranthaceae) ın Southern Algerian Sahara. Pak J Bot 50(1):313–320

Mukherjee K, Paul P, Banerjee ER (2014) Free radical scavenging activities of date palm (Phoenix sylvestris) fruit extracts. Nat Prod Chem Res 2(151):104172

Müller K, Borsch T (2005) Multiple origins of a unique pollen feature: stellate pore ornamentation in Amaranthaceae. Grana 44:266–281

Munien P, Naidoo Y, Naidoo G (2015) Micromorphology, histochemistry and ultrastructure of the foliar trichomes of *Withania somnifera* (L.) Dunal (Solanaceae). Planta 242(5):1107–1122

Mutlag SH, Ismael SH, Hassan AF, Abbas RF (2017) Genotoxic effect of ethyl acetate fraction of *Cressa cretica* on chromosomal aberration on bone marrow cells and spleen cells in mice. Int J Pharm Sci Rev Res 43(2):220–223

Nadkarni KM (1996) Indian materia medica: with ayurvedic, unani-tibbi, siddha, allopathic, homeopathic, naturopathic and home remedies, appendices and indexes, vol 2. Popular Prakashan Private Ltd, Ramdas Bhatkal

Nafea E (2017) Nutritive values of some wetland plants in the Deltaic Mediterranean Coast of Egypt. Egypt J Bot 57(1):1–10

Nafees M, Bukhari MA, Aslam MN et al (2019) Present status and future prospects of endangered Salvadora species: a review. J Glob Innov Agric Soc Sci 7:39–46

Naija DS, Bouzidi A, Boussaada O et al (2014) The antioxidant and free-radical scavenging activities of *Tamarix boveana* and *Suaeda fruticosa* fractions and related active compound. Eur Sci J 10(18):201–219

Najmi AK, Pillai KK, Pal SN, Aqil M (2004) Free radical scavenging and hepatoprotective activity of jigrine against galactosamine induced hepatopathy in rats. J Ethnopharmacol 97:521–525

Naresh V, Singh S, Sharma SK (2013) Antibacterial activity of stem bark of *Salvadora oleoides* Decne. Int J Pharmaco Phytochem Res 5:76–78

Nasernakhaei F, Zahraei M (2021) *Halocnemum strobilaceum* (Pall.) M. Bieb.: a review of its botany, phytochemistry, pharmacology and ethnobotany. J Med Plants 20(80):1–12

National Academy of Sciences (1980) Firewood crops: shrub and tree species for energy production. National Academy of Sciences, Washington, DC

National Research Council (2006) Lost crops of Africa: volume II: vegetables, vol 2. National Academies Press

Naveen G, Kumar A, Tyagi M et al (2015) Morphology of *Withania somnifera* (distribution, morphology, phytosociology of *Withania somnifera* L. Dunal). Int J Cur Res 1(7):164–173

Nazar S, Hussain MA, Khan A et al (2020) *Capparis decidua* Edgew (Forssk.): a comprehensive review of its traditional uses, phytochemistry, pharmacology and nutrapharmaceutical potential. Arab J Chem 13(1):1901–1916

Nazish M, Zafar M, Ahmad M et al (2019) Palyno-morphological investigations of halophytic taxa of Amaranthaceae through SEM from salt range of Northern Punjab, Pakistan. Microsc Res Tech 82(3):304–316

Nazish M, Ahmad M, Ullah R et al (2020) Taxonomic implications of leaf epidermis in halophytes of Amaranthaceae from Salt Range of Punjab, Pakistan. Plant Biosyst Int J Deal Aspects Plant Biol 2020:1–12

Nichita C, Neagu G, Cucu A et al (2016) Antioxidative properties of *Plantago lanceolata* L. extracts evaluated by chemiluminescence method. AgroLife Sci J 5(2):95–102

Nidavani RB, Mahalakhshmi AM, Shalawadi M (2014) Potent ulcer protective action of anti-inflammatory herbs: a short review. W J Pharm Res 3:2057–2066

Nikolova A, Vassilev A (2011) A study on *Tribulus terrestris* L. anatomy and ecological adaptation. Biotechnol Biotechnol Equip 25(2):2369–2372

Nima ZA-M, Jabier MS, Wagi RI, Hussain HAAK (2008) Extraction, identification and antibacterial activity of Cyperus oil from Iraqi *C. rotundus*. Eng Technol J 26(10):1156–1163

Noorbakhsh SN, Ghahraman A, Atar F, Mahdigholi K (2008) Leaf anatomy of *Artemisia* (Asteraceae) in Iran and its taxonomic implications. Iran J Bot 14(1):54–69

Oh GS, Yoon J, Lee GG et al (2015) The Hexane fraction of *Cyperus rotundus* prevents nonalcoholic fatty liver disease through the inhibition of liver X receptor alpha-mediated activation of sterol regulatory element binding protein-1c. Am J Chin Med 43(3):477–494

Olama HY, Shehata MN (1993) Phytosociological studies on the vegetation of Ayoum Musa area, south west Sinai. Desert Inst Bull Egypt 43:201–220

Onoja OJ (2016) Morpho-anatomical study on *Cleome viscosa* L. (Cleomaceae). J Pharmacognosy Phytochem 5(4):13

Osman A, Abdel Khalik KN (2005) Palynological study on some species of Convolvulaceae and its taxonomic significance. Taeckholmia 25(1):47–60

Othman A, Amen Y, Matsumoto M et al (2021) Bassiamide A, a new alkaloid from xero-halophyte *Bassia indica* Wight. Nat Prod Res 2021:1–9

Ozcan M (2005) Mineral composition of different parts of *Capparis ovata*. Heywood growingwild in Turkey. J Med Food 8(3):405–407

Ozturk M, Altay V, Gucel S, Guvensen A (2014) Halophytes in the east mediterranean-their medicinal and other economical values. In: Sabkha ecosystems: volume IV: cash crop halophyte and biodiversity conservation tasks for vegetation science, vol 47. Springer Netherlands. Springer Science+Business Media, Dordrecht, pp 247–272

Ozturk M, Altay V, Guvensen A (2019) Sustainable use of halophytic taxa as food and fodder – an important genetic resource in Southwest Asia. In: Hasanuzzaman M, Nahar K, Öztürk M (eds) Ecophysiology, abiotic stress responses and utilization of halophytes, vol 11. Springer, pp 235–257

Ozturk M, Altay V, Güvensen A (2021) *Portulaca oleracea*: a vegetable from saline habitats. In: Grigore MN (ed) Handbook of halophytes: from molecules to ecosystems towards biosaline agriculture. Springer, pp 2319–2332

Padalia H, Chanda S (2017) Characterization, antifungal and cytotoxic evaluation of green synthesized zinc oxide nanoparticles using Ziziphus nummularia leaf extract. Artif Cells Nanomed Biotechnol 45(8):1751–1761

Pal K, Rahaman CH (2014) Studies on foliar epidermal micromorphology, vegetative anatomy and xylem elements of four members of Portulacaceae. Int J Curr Res 6(2):4968–4975

Palmer J, Lally TR (2011) Amaranthaceae (version 1). In: Kellermann J (ed) Flora of South Australia, 5th edn). 42 pp. State Herbarium of South Australia, Adelaide

Panday HP (2004) Seed fume of *Solanum surattense*: a traditional panacea for teeth and gums. Indian J Tradit Knowl 3(2):206–207

Pandey AN, Rokad MV (1992) Sand dune stabilization: an investigation in the Thar Desert. India J Arid Environ 22:287–292

Pandey A, Singh R, Radhamani J, Bhandari DC (2010) Exploring the potential of *Ziziphus nummularia* (Burm. f.) Wight et Arn. from drier regions of India. Genet Resour Crop Evol 57(6):929–936
Pandey VC, Bajpai O, Pandey DN, Singh N (2015) *Saccharum spontaneum*: an underutilized tall grass for revegetation and restoration programs. Genet Resour Crop Evol 62(3):443–450
Pandey RK, Shukla SS, Jain A et al (2018) Evaluation of comparative immunomodulatory potential of *Solanum xanthocarpum* root and fruits on experimental animal. Indian J Pharm Educ Res 52(4 Suppl 2):237–245
Parmar S, Gangwal A, Sheth N (2010) *Solanum xanthocarpum* (Yellow Berried Night Shade): a review. Der Pharmacia Lett 2(4):373–383
Patil PV, Taware SP, Kulkarni DK (2014) Traditional knowledge of broom preparation from Bhor and Mahad region of western Maharashtra, India. Biosci Discov 5(2):218–220
Peachey LE, Pinchbeck GL, Matthews JB et al (2015) An evidence-based approach to the evaluation of ethnoveterinary medicines against strongyle nematodes of equids. Vet Parasitol 210(1/2):40–52
Pen-Mouratov S, Rakhimbaev M, Steinberger Y (2003) Seasonal and spatial variation in nematode communities in a Negev Desert ecosystem. J Nematol 35:157–166
Pereira CG, Custódio L, Rodrigues MJ et al (2017) Do potencial antioxidante e perfil fitoquímico do *Plantago coronopus*. Braz J Biol 77:632–641
Perveen A, Qaiser M (1996) Pollen Flora of Pakistan-VI. Salvadoraceae. Pakistan J Bot 28(2):151–154
Perveen A, Qaiser M (1998) Pollen Flora of Pakistan-XI. Leguminosae (Subfamily: Mimosoideae). Turk J Bot 22(3):151–156
Perveen A, Qaiser M (2001) Pollen flora of Pakistan-XXXI Capparidaceae. Turk J Bot 25(6):389–395
Perveen A, Qaiser M (2005) Pollen flora of Pakistan-XLIV. Rhamnaceae. Pak J Bot 37(2):195
Perveen A, Qaiser M (2006) Pollen flora of Pakistan-XLVIX. Zygophyllaceae. Pak J Bot 38(2):225
Perveen A, Qaiser M (2008) Pollen flora of Pakistan-LVI. Cucurbitaceae. Pak J Bot 40(1):9
Petropoulos SA, Karkanis A, Martins N, Ferreira IC (2018) Halophytic herbs of the Mediterranean basin: an alternative approach to health. Food Chem Toxicol 114:155–169
Phondani PC, Bhatt A, Elsarrag E, Alhorr YM (2015) Seed germination and growth performance of *Aerva javanica* (Burm. f.) Juss ex Schult. J Appl Res Med Aromat Plants 2(4):195–199
Pingale SS (2013) Evaluation of acute toxicity for *Solanum xanthocarpum* fruits. BioMedRx 1(3):330–332
Pirzada AM, Ali HH, Naeem M et al (2015) *Cyperus rotundus* L.: traditional uses, phytochemistry, and pharmacological activities. J Ethnopharmacol 174:540–560
Placines C, Castañeda-Loaiza V, João Rodrigues M et al (2020) Phenolic profile, toxicity, enzyme inhibition, in silico studies, and antioxidant properties of *Cakile maritima* Scop. (Brassicaceae) from southern Portugal. Plants 9(2):142
Pol M, Schmidtke K, Lewandowska S (2021) *Plantago lanceolata*-An overview of its agronomically and healing valuable features. Open Agric 6(1):479–488
Poonia A, Upadhayay A (2015) *Chenopodium album* Linn: review of nutritive value and biological properties. J Food Sci Technol (Mysore) 52(7):3977–3985
Pourabdollah Kaleybar P (2021) Phytochemical study and anti-cancer effects of *Halocnemum strobilaceum*. Tabriz University of Medical Sciences, School of Pharmacy, Tabriz
Priyashree S, Jha S, Pattanayak SP (2010) A review on *Cressa cretica* Linn.: a halophytic plant. Pharmacogn Rev 4(8):161
Punjani BL (2002) Ethnobotanical aspects of some plants of Aravalli hills in north Gujarat. Anc Sci Life 21(4):268
Puratuchikody A, Nithya DC, Nagalakshmi G (2006) Wound healing activity of Cyperus rotundus Linn. Indian J Pharm Sci 68:97–101
Qaiser M, Perveen A (2004) Pollen Flora of Pakistan-XXXVII. Tamaricaceae. Pakistan J Bot 36(1):1–18

Qari SH, Tawfik E, Hammad I (2019) Morphological, cytological, physiological and genetic studies of *Bassia indica* (Amaranthaceae). Gene Conserve 18(73):1–9
Qasim M (2014) Antioxidant properties of medicinal plants from coastal Pakistan. Ph. D. thesis, Institute of Sustainable Halophyte Utilization (ISHU), University of Karachi, Pakistan
Qasim M, Abideen Z, Adnan MY et al (2014) Traditional ethnobotanical uses of medicinal plants from coastal areas. J Coast Life Med 2(1):22–30
Quattrocchi U (2006) CRC world dictionary of grasses: common names, scientific names, eponyms, synonyms, and etymology. CRC Press, Taylor & Francis, Boca Raton
Quattrocchi U (2012) CRC world dictionary of medicinal and poisonous plants: common names, scientific names, eponyms, synonyms, and etymology. CRC Press, London
Qureshi R, Waheed A, Arshad M, Umbreen T (2009) Medico-ethnobotanical inventory of tehsil Chakwal, Pakistan. Pak J Bot 41(2):529–538
Qureshi R, Bhatti GR, Memon RA (2010) Ethnomedicinal uses of herbs from northern part of Nara desert, Pakistan. Pak J Bot 42(2):839–851
Rahamoz-Haghighi S, Bagheri K, Sharafi A, Danafar H (2021) Establishment and elicitation of transgenic root culture of *Plantago lanceolata* and evaluation of its anti-bacterial and cytotoxicity activity. Prep Biochem Biotechnol 51(3):207–224
Rahman MT, Ahmed M, Alimuzzaman M, Shilpi JA (2003) Antinociceptive activity of the aerial parts of *Solanum xanthocarpum*. Fitoterapia 74(1–2):119–121
Rajesh K, Singh Manish K, Avinash BK (2013) Ethnobotany of Tharus of Dudhwa national park, India. Mintage J Pharmaceut Med Sci 2(1):6–11
Rajput KS, Patil VS, Rao KS (2014) Stem anatomy of the dwarf subshrub *Cressa cretica* L. (Convolvulaceae). Flora Morphol Distribut Funct Ecol Plants 209(8):408–413
Ram H, Kumar A, Sharma SK et al (2012) Meiotic studies in *Withania somnifera* (L) Dunal.: a threatened medicinal herb of Indian Thar Desert. Am J Plant Sci 3:185–189
Rani S, Chaudhary S, Singh P et al (2011) *Cressa cretica* Linn: an important medicinal plant. A review on its traditional uses, phytochemical and pharmacological properties. J Nat Prod Plant Resour 1(1):91–100
Rashid AA, Perveen A (2014) Pollen morphology of some native and cultivated species of the genus *Phoenix* L. from Pakistan and Kashmir. Int J Biol Biotechnol 11(4):611–615
Rathee S, Mogla OP, Sardana S et al (2009) Phytochemical evaluation of stem of *Capparis decidua* (Forsk) Edgew. Pharm J 1:75–81
Rather LJ, Shahil-ul I, Mohammad F (2015) *Acacia nilotica* (L.): a review of its traditional uses, phytochemistry, and pharmacology. Sustain Chem Pharm 2:12–30
Rathore M (2009) Nutrient content of important fruit trees from arid zone of Rajasthan. J Hortic For 1(7):103–108
Razek MA, Moussa A, El-Shanawany Mand Singab A (2019) Comparative chemical and biological study of roots and aerial parts of *Halocnemum strobilaceum* growing wildly in Egypt. J Pharm Sci Res 11(9):3289–3296
Reddy KN, Naik CP, Reddy CS, Raju VS (2008) Plants used in traditional handicrafts in north eastern Andhra Pradesh. Indian J Tradit Knowl 7:162–165
Rekha R, Kumar SS (2014) Ethnobotanical plants used by the Malayali tribes in Yercaud hills of Eastern Ghats, Salem district, Tamil Nadu, India. Glob J Res Med Plants Indigenous Med 3(6):243
Rizwan M, Singh M, Mitra CK, Morve RK (2014) Ecofriendly application of nanomaterials: nanobioremediation. J Nanopart 2014:431787
Robertson S, Narayanan N, Deattu N, Nargis NR (2010) Comparative anatomical features of *Prosopis cineraria* (L.) Druce and *Prosopis juliflora* (Sw.) DC (Mimosaceae). Int J 2010:275–285
Rodrigues MJ, Gangadhar KN, Vizetto-Duarte C et al (2014) Maritime halophyte species from southern Portugal as sources of bioactive molecules. Mar Drugs 12:2228–2244

Rodriguez Zaragoza S, Mayzlish E, Steinberger Y (2005) Vertical distribution of the free-living amoeba population in soil under desert shrubs in the Negev Desert, Israel. Appl Environ Microbiol 71:2053–2060
Rudrappa T, Choi YS, Levia DF et al (2009) *Phragmites australis* root secreted phytotoxin undergoes photo-degradation to execute severe phytotoxicity. Plant Signal Behav 4(6):506–513
Sabnis TS (1921) The physiological anatomy of the plants of the Indian desert. J Indian Bot II(8–9):217–235
Sadou H, Sabo H, Alma MM et al (2007) Chemical content of the seeds and physico-chemical characteristic of the seed oils from *Citrullus colocynthis, Coccinia grandis, Cucumis metuliferus* and *Cucumis prophetarum* of Niger. Bull Chem Soc Ethiop 21(3):323–330
Said M (1984) Potential of herbal medicines in modern medical therapy. Anc Sci Life 4:36–47
Said O, Fulder S, Khalil K et al (2002) Maintaining a physiological blood glucose level with `Glucolevel', a combination of four antidiabetes plants used in the traditional Arab herbal medicine. Evid-Based Complement Altern Med 5:421–428
Saleem H, Ahmad I, Zengin G et al (2020) Comparative secondary metabolites profiling and biological activities of aerial, stem and root parts of *Salvadora oleoides* Decne (Salvadoraceae). Nat Prod Res 34(23):3373–3377
Saleem H, Khurshid U, Sarfraz M et al (2021) A comprehensive phytochemical, biological, toxicological and molecular docking evaluation of *Suaeda fruticosa* (L.) Forssk.: an edible halophyte medicinal plant. Food Chem Toxicol 154:112348
Samejo MQ, Memon S, Bhanger MI, Khan KM (2012) Chemical composition of essential oils from *Alhagi maurorum*. Chem Nat Compd 48(5):898–900
Samir D, Manel A, Abir H (2019) Phytochemical analysis and antioxidant property of rhizome extracts aqueous of *Phragmites australis* in alloxan diabetic rats. Asian J Pharm Technol 9(4):249–252
Sathya M, Kokilavani R (2013) Phytochemical screening and in vitro antioxidant activity of *Saccharum spontaneum* Linn. Int J Pharm Sci Rev Res 18(1):75–79
Satyanarayana T, Anjana A, Vijetha P (2008) PHCOG REV.: plant review phytochemical and pharmacological review of some Indian *Capparis* species. Pharmacognosy Rev [Phcog Rev] Suppl 2(4):36–45
Savithramma N, Sulochana C, Rao KN (2007) Ethnobotanical survey of plants used to treat asthma in Andhra Pradesh, India. J Ethnopharmacol 113(1):54–61
Schafferman D, Beharav A, Shabelsky E, Yaniv Z (1998) Evaluation of *Citrullus colocynthis*, a desert plant native in Israel, as a potential source of edible oil. J Arid Environ 40(4):431–439
Scordia D, Cosentino SL, Jeffries TW (2010) Second generation bioethanol production from *Saccharum spontaneum* L. ssp. *aegyptiacum* (Willd.) Hack. Bioresour Technol 101:5358–5365
Sekina MA, Moore PD (1995) Morphological studies of the pollen grains of the semi-arid region of Egypt. Flora 190(2):115–133
Semerdjieva I, Yankova-Tsvetkova E, Baldjiev G, Yurukova-Grancharova P (2011) Pollen and seed morphology of *Tribulus terrestris* L. (Zygophyllaceae). Biotechnol Biotechnol Equip 25(2):2379–2382
Senthil K, Wasnik NG, Kim Y, Yanj D (2009) Generation and analysis of expressed sequence tags from leaf and root of *Withania somnifera* (Ashwagandha). Mol Biol Rep 37:893–902
Shah M, Hussain F (2012) Ethnomedicinal plant wealth of Mastuj valley, Hindukush range, District Chitral, Pakistan. J Med Plants Res 6:4328–4337
Shah M, Sayyed A, Sherwani SK (2014) Ethno ecological study of fuel and timber wood species of the most important market of Khyber Pukhtoonkhwa, Pakistan. World Appl Sci J 31(4):420–426
Shah MZ, Guan ZH, Din AU et al (2021) Synthesis of silver nanoparticles using *Plantago lanceolata* extract and assessing their antibacterial and antioxidant activities. Sci Rep 11(1):1–14
Shaheen S, Ahmad M, Khan F et al (2011) Morpho-playnological and foliar epidermal anatomy of genus *Cenchrus* L. J Med Plants Res 5(16):3796–3802

Shajib M, Akter S, Ahmed T, Imam MZ (2015) Antinociceptive and neuropharmacological activities of methanol extract of Phoenix sylvestris fruit pulp. Front Pharmacol 6:212
Shaltout KH, El-Beheiry MA (2000) Demography of *Bassia indica* in the Nile Delta region, Egypt. Flora 195:392–397
Shaltout KH, Mady MA (1996) Analysis of raudhas vegetation in central Saudi Arabia. J Arid Environ 34:441–454
Shamso E, Toshiyuki F (2012) The Pollen Flora of Faiyum, Egypt I-Archichlamydeae. Taeckholmia 32(1):1–40
Sharma N, Sharma AK, Zafar R (2010) Kantikari: a prickly medicinal weed ~ ecosensorium. J Phytol Res 9(1):13–17
Sharma DC, Shukla R, Ali J et al (2016) Phytochemical evaluation, antioxidant assay, antibacterial activity and determination of cell viability (j774 and thp1 alpha cell lines) of P. sylvestris leaf crude and methanol purified fractions. EXCLI J 15:85–94
Shinwari ZK, Watanabe T, Rehman M, Youshikawa T (2006) A pictorial guide to medicinal plants of Pakistan. Kohat University of Science &Technology, Kohat
Shinwari ZK, Ahmad N, Ahmad I et al (2019) Biochemical screening of crude extract and its derived fractions obtained from *Calligonum polygonoieds* and *Typha latifolia*. Pak J Bot 51(3):1107–1111
Shomer-Ilan A, Nissenbaum A, Waisel Y (1981) Photosynthetic pathways and the ecological distribution of the Chenopodiaceae in Israel. Oecologia 48:244–248
Sikarwar I, Dey YN, Wanjari MM et al (2017) *Chenopodium album* Linn. leaves prevent ethylene glycol-induced urolithiasis in rats. J Ethnopharmacol 195:275–282
Silva VDA, Pitanga BP, Nascimento RP et al (2013) Juliprosopine and Juliprosine from *Prosopis juliflora* leaves induce mitochondrial damage and cytoplasmic vacuolation on cocultured glial cells and neurons. Chem Res Toxicol 26(12):1810–1820
Singh S, Swapnil S (2011) Antibacterial properties of alkaloid rich fractions obtained from various parts of *Prosopis juliflora*. Int J Pharma Sci Res 2:114–120
Singh Y, Wadhwani AM, Johri BM (1990) Dictionary of economic plants of India. Indian Council of Agricultural Research, New Delhi
Singh BK, Gahoi R, Sonkar A (2010a) A quality assessment and phytochemical screening of selected region of *Withania somnifera* Dunal. Int J Pharmaceut Sci Res 1:73–77
Singh P, Shivhare Y, Singhai AK, Sharma A (2010b) Pharmacological and phytochemical profile of *Chenopodium album* Linn. Res J Pharm Technol 3(4):960–963
Singh JP, Mathur BK, Rathore VS, Beniwal RK (2012) Weeds as a source of fodder in hot arid zone – a review. In: Feeding and management of livestock during drought and scarcity, vol 124. Scientific Publishers
Sivapalan SR (2013) Medicinal uses and pharmacological activities of Cyperus rotundus Linn – a review. Int J Sci Res Publ 3(5):1–8
Slama K, Boumendjel M, Taibi F et al (2020) *Atriplex halimus* aqueous extract abrogates carbon tetrachloride-induced hepatotoxicity by modulating biochemical and histological changes in rats. Arch Physiol Biochem 126(1):49–60
Smaoui A, Barhoumi Z, Rabhi M, Abdelly C (2011) Localization of potential ion transport pathways in vesicular trichome cells of *Atriplex halimus* L. Protoplasma 248:363–372
Somasiri SC, Kenyon PR, Morel PCH et al (2020) Selection by lambs grazing plantain (*Plantago lanceolata* L.), chicory (*Cichorium intybus* L.), white clover (*Trifolium repens* L.), red clover (*Trifolium pratense* L.) and perennial ryegrass (*Lolium perenne* L.) across seasons. Animals 10:2292
Sosam AA, Al-Mayyahi TF (2018) Pollen morphology of some species of *Cyperus* L. and *Bolboschoneus* is growing in Diwaniyah river. Res J Pharm Technol 11(4):1573–1579
Sultana A, Rahman K (2013) *Portulaca oleracea* Linn. A global Panacea with ethno-medicinal and pharmacological potential. Int. J Pharm Pharm Sci 5(2):33–39
Talasaz AH, Abbasi MR, Abkhiz S, Dashti-Khavidaki S (2010) *Tribulus terrestris*-induced severe nephrotoxicity in a young healthy male. Nephrol Dial Transplant 25(11):3792–3793

Tawfik MM, Galal B, Nafie MS et al (2021) Cytotoxic, apoptotic activities and chemical profiling of dimorphic forms of Egyptian halophyte *Cakile maritima* Scop. J Biomol Struct Dyn 2021:1–14
Tekuri SK, Pasupuleti SK, Konidala KK (2019) Phytochemical and pharmacological activities of *Solanum surattense* Burm. f. – a review. J Appl Pharmaceut Sci 9(03):126–136
Tene V, Malagón O, Finzi PV et al (2007) An ethnobotanical survey of medicinal plants used in Loja and Zamora Chinchipe, Ecuador. J Ethnopharmacol 111:63–81
Thalen DCP (1979) Ecology and utilization of desert shrub rangelands in Iraq. Dr. W. Junk, BV Publishers, The Hague
Tjitrosoedirdjo SS, Sastroutomo SS (1986) The uptake of some heavy metals by *Typha angustifolia*. Biotrop Spec Pub 24:93–99
Toma C, Grigore MN, Afemei M, Stanescu IE (2010) Histo-anatomical considerations on some Romanian *Inula* L. species, with pharmacological action. Analele Stiintifice ale Universitatii "Al. I. Cuza" din Iasi 56(1):5
Turki Z, El-Shayeb F, Shehata F (2006) Taxonomic studies in the Camphorosmeae (Chenopodiaceae) in Egypt. 1. Subtribe Kochiinae (*Bassia, Kochia* and *Chenolea*). Fl Medit 16:275–294
Ullah M, Khan MU, Mahmood A et al (2013) An ethnobotanical survey of indigenous medicinal plants in Wana district south Waziristan agency, Pakistan. J Ethnopharmacol 150(3):918–924
Ullah I, Ahmad M, Jabeen A et al (2021) Palyno-morphological characterization of selected allergenic taxa of family Poaceae from Islamabad-Pakistan using microscopic techniques. Microsc Res Tech 84:2544–2558
Umadevi M, Rajeswari R, Rahale CS et al (2012) Traditional and medicinal uses of *Withania somnifera*. Pharma Innov 1(9):102
Umerie SC, Ezeuzo HO (2000) Physicochemical characterization and utilization of Cyperus rotundus starch. Bioresour Technol 72:193–196
Umesh MK, Sanjeevkumar CB, Hanumantappa BNRL, Ramesh L (2014) Evaluation of in vitro anti-thrombolytic activity and cytotoxicity potential of *Typha angustifolia* L leaves extracts. Int J Pharm Pharm Sci 6(5):81–85
Verma S (2016) *Salvadora oleoides* (meethi-jal) Salvadoraceae: phyto-chemical study. Int J Pharm Chem Biol Sci 6(4):1
Waaris HM, Ahmad S, Anjum S, Alam K (2014) Ethnobotanical Studies of Dicotyledonous Plants of Lal Suhanra National Park, Bahawalpur. Int J Sci Res 3:2452–2460
Walker DJ, Lutts S, Sánchez-García M, Correal E (2014) *Atriplex halimus* L.: its biology and uses. J Arid Environ 100:111–121
Wang YM, Wang JB, Luo Da Jia JF (2001) Regeneration of plants from callus cultures of roots induced by *Agrobacterium rhizogenes* on *Alhagi pseudalhagi*. Cell Res 11:279–284
Wapakala W (1966) A note on the persistence of mulch grasses. Kenya Coffee 31:111–112
Wariss HM, Ahmad S, Alam K et al (2017) Floristic inventory of Baghdad-ul-jadeed campus, the Islamia university of Bahawalpur, Pakistan. South Asian J Life Sci 5(1):5–18
Wasim MA, Naz N (2020) Anatomical adaptations of tolerance to salt stress in *Cenchrus ciliaris* L., a saline desert grass. J Anim Plant Sci 30(6):1548–1566
Watt JM, Breyer-Brandwijk MG (1962) The medicinal and poisonous plants of Southern and Eastern Africa. E & S Livingstone Ltd, Edinburgh and London
Wazir R, Muhammad A, Subhan M et al (2014) Morpho-anatomical features of weed flora of rainfed maize fields in Mir Ali, North Waziristan Agency, Pakistan. Pakistan J Weed Sci Res 20(3):385–403
Weber E (2003) Invasive plant species of the world: a reference guide to environmental weeds. CAB International, Wallingford
WHO (World Health Organization) (2013) Medicinal plants in Mongolia. World Health Organization Regional Office for the Western Pacific
Wickens GE, Field DV, Goodin JR (eds) (2012) Plants for arid lands: proceedings of the Kew international conference on economic plants for arid lands held in the Jodrell Laboratory, Royal Botanic Gardens, Kew, England, 23–27 July 1984. Springer Science & Business Media

Wiersema JH, León B (1999) World economic plants: a standard reference. CRC Press, Boca Raton
Wills GD, Briscoe GA (1970) Anatomy of purple nutsedge. Weed Sci 18(5):631–635
Windadri FI (2001) *Cleome viscosa* L. Record from Proseabase. PROSEA (Plant Resources of South-East Asia) Foundation, Bogor
www.cabi.org. Accessed 15 Apr 2022
www.efloras.org. Accessed 15 Apr 2022
www.indiabiodiversity.org. Accessed 15 Apr 2022
www.keyserver.lucidcentral.org. Accessed 15 Apr 2022
www.paldat.org. Accessed 15 Apr 2022
www.powo.science.kew.org. Accessed 15 Apr 2022
www.tropical.theferns.info. Accessed 15 Apr 2022
www.tropicos.org. Accessed 15 Apr 2022
www.worldagroforestry.org. Accessed 15 Apr 2022
www.worldfloraonline.org. Accessed 15 Apr 2022
Xu D, Zheng J, Zhou Y et al (2017) Ultrasound-assisted extraction of natural antioxidants from the flower of *Limonium sinuatum*: optimization and comparison with conventional methods. Food Chem 217:552–559
Yanbin D, Yancheng J, Jian L (1998) The xeromorphic and saline morphic structure of leaves and assimilating branches in ten Chenopodiaceae species in Xinjiang. Chinese J Plant Ecol 22(2):164
Younos C, Fleurentin J, Notter D et al (1987) Repertory of drugs and medicinal plants used in traditional medicine of Afghanistan. J Ethnopharmacol 20(3):245–290
Yousaf MZ (2014) Ethnobotanical studies of Khushab District, Punjab, Pakistan. Eur J Appl Sci Technol 1(4):110–119
Youssef RS (2013) Medicinal and non-medicinal uses of some plants found in the middle region of Saudi Arabia. J Med Plants Res 7(34):2501–2513
Yu J, Lei G, Cai L, Zou Y (2004) Chemical composition of C. rotundus extract. J Phytochem 65:881–889
Yurukova LD, Kochev KK (1993) Energy content and storage in the biomass of hydrophytes in Bulgaria. Arch Hydrobiol 127(4):485–495
Zaidi SF, Muhammad JS, Shahryar S et al (2012) Anti-inflammatory and cytoprotective effects of selected Pakistani medicinal plants in *Helicobacter pylori*-infected gastric epithelial cells. J Ethnopharmacol 141:403–410
Zaier MM, Ciudad-Mulero M, Cámara M et al (2020) Revalorization of Tunisian wild Amaranthaceae halophytes: nutritional composition variation at two different phenotypes stages. J Food Compost Anal 89:103463
Zali SH, Tahmasb R (2016) Medicinal plants of Farashband tribe's winter pastures and their traditional uses. J Adv Health Med Sci 2(1):18–27
Zarinkamar F (2006) Foliar anatomy of Chenopodiaceae family and xerophytes adaptation. Iranian J Bot 11(2):175–183
Zaurov DE, Belolipov IV, Kurmukov AG et al (2013) The medicinal plants of Uzbekistan and Kyrgyzstan. In: Eisenman S, Zaurov D, Struwe L (eds) Medicinal plants of Central Asia: Uzbekistan and Kyrgyzstan. Springer, New York. https://doi.org/10.1007/978-1-4614-3912-7_5
Zhou YX, Xin HL, Rahman K et al (2015) *Portulaca oleracea* L.: a review of phytochemistry and pharmacological effects. Biomed Res Int 2015:Article ID 925631
Zhu M, Luk HH, Fung HS, Luk CT (1997) Cytoprotective effects of Cyperus rotundus against ethanol induced gastric ulceration in rats. Phytother Res 11:392–394
Zhu W, Du Y, Meng H et al (2017) A review of traditional pharmacological uses, phytochemistry, and pharmacological activities of *Tribulus terrestris*. Chem Cent J 11(1):1–16

Chapter 3
Phytochemistry and Biological Activity in the Halophytes

3.1 An Overview of the Phytochemical Diversity

3.1.1 Important Sources of Primary and Secondary Metabolites

Halophytes are highly specialized plants having specific morphological and physiological adaptations that allow them to survive under saline conditions. Their tolerance is determined by different factors, as they have variety of adaptations to maintain their ion homeostasis in such environments and they acquire or store water as well as protect their cells from a damage due to ROS "Reactive Oxygen Species" accumulation (Flowers and Colmer 2008; Shabala and Mackay 2011). These plants possess different types of biocompounds which are useful in view of their biological activities such as anti-microbial and antioxidant ones (Ksouri et al. 2012a). The biocompounds increase the nutraceutical value of such plant taxa.

Galvani (2007) has reported that different profitable ingredients such as fibers, oils, gums, resins, flavors and essential oils are found in the halophyte taxa, but the change in their metabolic profile depends on the species, genus, cultivar, saline condition, and growth stage of the species (Bernstein et al. 2010).

The plants included in this ecological group are a source of several primary and secondary metabolites which play an important role in overcoming oxidative stress under saline conditions (Bose et al. 2014). Environmental stress generates oxidative stress in the form of ROS, which can damage the cellular structure. The halophytes cope with this condition by increasing biosynthesis of metabolites which are osmotically active, antioxidants and proteins. A positive correlation has been observed between saline medium and increased production of secondary metabolites in halophytes (Buhmann and Papenbrock 2013). One of the typical examples is *Nitraria*

M. Ozturk et al., *Halophyte Plant Diversity and Public Health*,
https://doi.org/10.1007/978-3-031-21944-3_3

retusa which produces more flavonoid and phenol content under increased salt concentrations (Boughalleb and Denden 2011). An accumulation of the osmoprotectants such as polyphenols, proline, soluble sugars, inorganic ions, and glycine is evaluated as a defense strategy by these plants to deal with the environmental stresses (Patel et al. 2016).

Thousands of compounds have been reported in different studies on the plants and among these, especially considering their molecular structures, secondary metabolites show greater diversity than primary ones. The structural properties of secondary compounds reveal that these can generally be grouped chemically under different groups such as alkaloids, phenolics, acetogenins and terpenoids (Wink 2003; Stevanovic et al. 2019). The richness and diversity of secondary metabolites available in the plant kingdom represent a crucial biogenic resource for the discovery of potential new drugs. Very few studies are available on salt-tolerant plants which occupy ecologically special habitats on the planet and we are trying to make them wonder as a separate mysterious natural potential resource (Dajic Stevanovic et al. 2014; Stevanovic et al. 2019). Inspite of the limited studies, the halophytes have been recognized as sources of polyphenols, terpenes and many other potential secondary metabolites. Thus, the bioactive effects of these plants and their possible uses as drugs and therapeutics has gained a great impetus (Menzel and Lieth 1999).

As the world population is increasing day by day, a constant reduction in arable land is observed which is leading towards an increase in the area of saline soils. Salinity has now started becoming a global threat to food security but hope lies in the plants which occupy such habitats i.e., the halophytes. Such plants have great potential for producing nutraceuticals and they could be a rich source of food in saline-alkaline areas (Hameed et al. 2010). Maria et al. (2019) have reported at length the presence of fatty acids, cinnamic acid, acylglycerols, carboxylic acids, alkanes, alkenes, amino acids, carbohydrates, alcohols, ketones, flavonoids, aldehydes, alkaloids, terpenoids, polyphenols, tocopherols, and stilbenoids in halophytic grasses.

Primary metabolites are the ones which play a direct role in growth, development, and differentiation. These are generally synthesized directly by sunlight and have a physiological function in the organism, being major constituent of organisms or cells (Bernal et al. 2011; Arya et al. 2019). The carbohydrates and their derivatives, lipids, certain amino acids, and related compounds are the primary metabolites (Ksouri et al. 2012a; Arya et al. 2019). Carbohydrates (especially polysaccharides) act as structural components in living beings and are important molecules for energy storage. These molecules also show important bioactivity features such as prebiotics, antioxidants, and anti-inflammators (Bernal et al. 2011). The halophytes are drawing an attention as a natural source of compounds in terms of lipids – the natural molecules which contain a rich group, especially fatty acids, waxes, sterols, tocopherols, mono-, di- and triacylglycerols, phospholipids and glycolipids (Parida et al. 2019). The presence of these molecules in this group of plants adds to their commercial value for many economic applications such as medicinal, cosmetic, pharmaceutical, food and nutraceutical ones, and also create an alternative potential source for biotechnological applications (Parida et al. 2019). Potential bioactivity (especially antioxidant and antimicrobial) studies in many lipid molecules of halophytes have also been carried out by different researchers (de Jesus Cortés-Sánchez et al. 2013; Rodrigues et al. 2014; Parida et al. 2019). Interestingly, in the seeds of various halophyte species, presence of high amounts of polyunsaturated fatty acids shows that these can be used as a quality oil source, important in human nutrition (Ksouri et al. 2012a). In fact, in several studies undertaken by different researchers, it has been reported that these oils contain 70–80% unsaturated enough good quality edible oil and have higher nutritional value compared to traditionally used vegetable oils (Stuchlík and Žák 2002; Ksouri et al. 2012a, 2012b; Qasim et al. 2011).

Acylglycerols in halophytes have great nutritional importance as it is an edible oil source for industries and for the treatment of various diseases (Cagliari et al. 2011). The benzoic acids are reported as being the building blocks of primary metabolites in halophytes (Widhalm and Dudareva 2015). Carboxylic acids and cinnamic acid in halophytes highlight the fact that halophytes can prove beneficial in human diet. The carbohydrates are the source of energy and sugars add flavor to the plants, their presence reveals that halophytes have a good potential as functional foods. These plants are capable of producing flavonoids and phenols against oxidative stress for their defense mechanisms under extreme saline conditions (Mohammed 2020).

Amino acids are important for maintaining the diet and are of great nutritional significance. They play an important role during osmotic stress by maintaining the homeostasis, and are therefore important in plant physiology. Some of the important amino acids in halophytes are lysine, histidine, leucine, valine, threonine, and methionine which reveals their importance as vegetal source (Fan et al. 1993; Flowers and Colmer 2008).

Proline plays a significant role in maintaining osmotic balance under the salt stress condition (Ozturk and Szaniawiski 1981; Ozturk et al. 1986, 2011; Turkyilmaz Unal et al. 2013; Mansour and Ali 2017). It also acts as a protein and membrane structure stabilizer and ROS scavenger (Rejeb et al. 2014). Glycine betaine is also associated with the osmotic adjustment and play important role in enzyme protection and membrane stabilization (Sakamoto and Murata 2000; Chen and Murata 2011). Trehalose protects halophytes under abiotic stress (Mostafa et al. 2015). The sugars such as fructose and glucose are also involved in the maintenance of osmotic potential under salt and drought stress (Jayakannan et al. 2013). The application of organic amines or polyamines have shown increased salinity tolerance (Minocha et al. 2014).

Salt tolerant plants are the main source of natural trace elements and antioxidants also (Qasim et al. 2011). Zinc, iron, and copper have been reported in halophytes in high concentrations from the region of high saline soil. Copper is an important constituent of an antioxidant that is the superoxide dismutase enzyme present in halophytes (Scheiber et al. 2014). The development stage of plants plays important role in the concentration or production rate of secondary metabolites. *Crithmum maritimum* and *Mesembryanthemum crystallinum* have shown the highest flavonoids and phenols concentrations during the flowering period (Jallali et al. 2012; Ahmad et al. 2013).

Stevanović et al. (2019) have reported different secondary metabolites in the halophytic members of various families such as Amaranthaceae, Apiaceae, Brassicaceae, Combretaceae, Euphorbiaceae, Fumariaceae, Lamiaceae, Plantaginaceae, Primulaceae, Saxifragaceae, Salvadoraceae, and Tamariaceae. The reported secondary metabolites are phenolic acids, flavonoids, aromatic acids, isoflavone glycosides, tannins, and coumarins. They have also reported fatty acids, saponins, carotenoids, essential oils, polyacetylenes, and alkaloids in halophytes (Ozturk et al. 2019; Raza et al. 2019; Shahrasbi et al. 2021; Khan et al. 2021; Caparrós et al. 2022; Younessi-Mehdikhanlou et al. 2022).

While the majority of bioactive compounds generally identified from halophytes are fatty acids and phenolic compounds, some lipid and glycoside derivatives have also been identified (Rodrigues et al. 2014; Plouguerné et al. 2014; Parida et al. 2019). Abdel-Aziz et al. (2016) have reported that mangrove species, which represent an ecologically special group among halophytes, have bioactive compounds rich in flavonoids, triterpenes, alkaloids, saponins, fatty acids, tannins, and quinones.

The hydroxyl group attached to the aromatic ring is the general structural characteristic of all phenolic derivatives. A large group of natural products with this structural fragment are the "natural aromatic compounds," showing high antioxidant properties depending on the nature and structure of the substitutions in the aromatic rings in the chemical structure of phenolic compounds (Bernal et al. 2011; Stanković et al. 2019). The phenolic compounds and flavonoids comprise all phenolic derivatives, systematized in these two groups. The phenols, phenolic acids, and condensation products of phenol – dephides, phenylpropane, tannins, and stilbene form the first group whereas while the flavonoids and anthocyanins form the other group (Crozier et al. 2006; Quideau et al. 2011; Stanković et al. 2019). Moreover, these compounds can be found in different parts of halophytes. Different amounts of phenolic compounds from plant sources change to some extent and this depends on their function in the plant itself, particularly their role in photosynthetic apparatus protection as well as protection against herbivores and pathogens, including other biotic and abiotic factors to which the plant is exposed during growth and development (Alonso-Amelot et al. 2004; Stanković et al. 2019). One of the common abiotic factors in saline habitats is the physiological drought, caused by the increased intensity of synthesis and accumulation of phenolic compounds to prevent the negative consequences of drought (Stanković et al. 2019). It has been reported that some halophytes are effective in human nutrition due to these chemical compounds, as flavor precursors. These plants also show different bioactiviies such as protection of blood sugar levels, and anti-inflammatory and anti-aging in humans (Pandey and Rizvi 2009; Parida et al. 2019). During the last two decades several authors have emphasized that halophytes are rich in phenolic molecules (Ksouri et al. 2009; Falleh et al. 2011; Jdey et al. 2017).

Other important natural molecules that exhibit different biological and pharmacological activities in the treatment of many diseases are alkaloids (Stöckigt et al. 2002). Different alkaloid compounds have been identified in many halophytes. Notable among these are natural alkaloid compounds such as quinolizidines, spiropiperidines, and quinazolines identified from different species of genus *Nitraria*, and salsolin and salsolidin identified from the *Salsola kali* plant (Wang et al. 2007; Sokolowska-Krzaczek et al. 2009).

In a phytochemical study conducted on some halophyte species, it has been reported that the presence of terpenoids including monoterpenes, sesquiterpenes and diterpenoids, makes significant contributions to the medicinal, functional, nutritional and industrial usefulness quality of related halophytes (Qian et al. 2017). Phytochemical compounds revealed in some of the halophytes to date are listed in Table 3.1.

Table 3.1 Some important phytochemical sources of primary and secondary metabolites, osmolytes, minerals, and amino acids in halophytes

Phytochemical compounds
13-earbolines
2,6-dimethyl phenol
2-methyl butyl glucosinolate
3-O-β-D-glucopyranoside
4-pentyl glucosinolate
8-Omethylretusin
acid polysaccharides
acylglycerol
aldehydes
aliphatic acids
alkaloids
alkanes
apigenin
ascorbic acid
avicularin
benzoic acid
benzoquinones
benzyl benzoate
beta-alanine
betacyanins
betaine
caffeic acid
calcium
camphor
carboxylic acid
carotenoids
catechin
catechin hydrate

(continued)

Table 3.1 (continued)

chalcones
chenopodine
chlorogenic acid
cinnamic acid
copper
coumarin
cyanidin-3-O-diglucoside
cyclitol
diterpenes
ellagic acid
epicatechin
epigallocatechin
epigallocatechin-3-o-gallate
essential oils
ethyl glucosinolate
eugenol
falcarindiol
fatty acid
fenchyl acetate
ferulic acid
flavone
flavonoid glycosides
flavonoids
flavonol
flavonol glycoside
fructose
furochromone
gallic acid
gallocatechin
germacrene D
glucosinolates
glucotropaeolin
glucuronoxylans
glycerol
glycine
glycinebetaine
glycosides
hexadecanoic acid docosane
histidine
hydroxyacetophenone
hydroxyl fatty acid
hyperoside
ısoflavone glycoside

(continued)

Table 3.1 (continued)

ısorhamnetin
ısorhamnetin 3-O-glucoside
inositol
inositol
iron
isochinolin
isoquercetin
juncunol
kaempferol
ketones
lactone
leucine
linoleic acid
lipids
lipopolysaccharide
luteolin
lysine
magnesium
maltose
mannitol
m-coumaric acid
methionine
methyl-4-hydroxybenzoate
monoterpenes
myo-inositol
myrctin
myricetin
narinenin 7-O-glucoside
naringin
n-propyl -3,4,5-trihydroxybenzoate
oleic acid
omega-3 polyunsaturated fatty acid
ononitol
patuletin
pelargonidin-3-O-diglucoside
Phenanthrenoids
phenolic acids
phenolic compounds
phenols
phloretin
phytoalexins
pinitol
pipecolate

(continued)

Table 3.1 (continued)

plumbocatechin A
polyacetylene
polyamines
polyphenol gallic
polyphenols
polyunsaturated fatty acids
potassium
proanthocyanidins
procatechuic
procyanidins
proline
propelargonidins
proteins
pyrocatechol
quercetin
quercetin 3-O-glucoside
quercetin dihydrate
quinazolines
quinolizidines
resorcinol
rhamnose
rosmarinic acid
rutin
salicylic acids
salsolidin
salsolin
saponins
scopoletin
sesquiterpenes
sodium
sorbitol
sorbitol
spinacetin
spiropiperidines
steroids
sterols
stilberoids
sucrose
sulfated flavonoids
sulfur glycosides
syringic acid
tannins
threonine

(continued)

Table 3.1 (continued)

trans3-hydroxycinnamic acid
trans-cinnamic acids
trehalose
tricin
tridecane
trigonelline
triterpenes
triterpenic ketone
triterpenoid saponins
tungtungmadic acid
tyramine
valine
vanillic acid
vitamin A
vitamin B
vitamin C
vitamin E
α humulene
α-terpinene
β-carotene

Modified from Ksouri et al. (2012a), Slama et al. (2015), Arya et al. (2019), Parida et al. (2019), Stevanovic et al. (2019), Stanković et al. (2019), Obón et al. (2021), Stanković and Jakovljević (2021)

3.2 Biological Activity

Although showing an ability to overcome different extreme ecological conditions, halophytes have a strong biological capacity because of their ability to produce secondary compounds, especially phenolic molecules, as a result of metabolic activities (Trabelsi et al. 2012; Stanković et al. 2019). The metabolites found in the halophytes have a great potential to exhibit many biological activities, especially antioxidant, antimicrobial, and anti-inflammatory. Such halophytic metabolites can represent rich alternative natural resources in the treatment of a multitude of diseases, popular in today's world, especially cancer and cardiovascular disorders (Ksouri et al. 2012a; Rodrigues et al. 2015; Stanković et al. 2019). During the last two decades, studies on several potential biological activities, especially antioxidant activity, of some halophytic species has been reported (Meot-Duros et al. 2008; Lee et al. 2011; Oueslati et al. 2012a, b; Trabelsi et al. 2012; Slama et al. 2015; Arya et al. 2019; Parida et al. 2019; Stevanovic et al. 2019; Stanković et al. 2019; Obón et al. 2021; Stanković and Jakovljević 2021).

3.2.1 Antioxidant Activity

Antioxidants are the natural biomolecules, special metabolites which inhibit the harmful activities caused by free radicals. They appear as biomolecules preventing the oxidation of substrates exposed to the oxidation process. The mechanism of action of these beneficial metabolites is based on their role as electron or hydrogen atom donors, as well as their ability to destroy the hydroperoxide of lipid molecules (Xu et al. 2017; Stevanovic et al. 2019).

Oxidative stress is the additional phenomenon of stress which has great impact on plants, as a consequence of hyperosmolarity under conditions of increased salinity, resulting in the appearance of reactive oxygen molecules, mainly hydrogen peroxide, hydroxyl radicals and superoxide anions (Xiong et al. 2002; Stevanovic et al. 2019; Garcia-Caparros et al. 2021).

According to Stevanovic et al. (2019), among the secondary metabolites, main metabolite polyphenol compounds are also effective in antioxidant activity. These authors have also emphasized the fact that qualitative compositions of these compounds in the halophytes generally varies depending on the taxonomic status of species together with the ecological conditions of habitat where it grows (Stevanovic et al. 2019).

As a rich source of natural antioxidants, phenolic compounds are the most prominent secondary metabolites. They possess multiple applications such as additives, functional food, and medical and pharmaceutical materials (Povichit et al. 2010; Stanković et al. 2019). The pharmacological studies have been conducted in recent years on some halophytes in an attempt to characterize their potential uses and the reports reveal that these plants comprise rich sources of natural antioxidants (Gourine et al. 2010; Ksouri et al. 2012a; Stanković et al. 2019).

3.2.2 Antimicrobial Activity

Many researchers have reported that halophytes have antimicrobial activity, because of the presence of important secondary metabolites in these such as phenolic acids, flavonoids, terpenoids, coumarins, and tannins (Stevanovic et al. 2019). As far as the mechanism of action is concerned, particularly phenolic acids causing selective loss of permeability of the bacterial membrane, flavonoids completely block some enzymes involved in the replication process of bacterial DNA, terpenoids disrupt the bacterial cell membrane, coumarins inhibit cellular respiration, tannins disrupt the biomolecular functioning and enzyme activity of bacterial membranes (Cowan 1999; Cushnie and Lamb 2005; Lou et al. 2011; Stevanovic et al. 2019). Moreover, as the habitats of halophytes vary, the antimicrobial effective features too show a different intensity depending on the density of the secondary metabolites present in the plant (Stevanovic et al. 2019).

3.2.3 Anticancer Activity

Some of the secondary plant metabolites are reported to affect uncontrolled cell division. However, the antitumor activity mechanism of such metabolites differs depending on the cell and tissue type. They often have the ability to induce apoptosis, inhibit angiogenesis, or alter the activity of enzymes involved in cell cycle regulation (Ren et al. 2003; Duangmano et al. 2010; Khoo et al. 2010; Stevanovic et al. 2019). There are few studies on the anticancer activity of halophyte metabolites, albeit limited ones (Kang et al. 2011; Ksouri et al. 2013; Wang et al. 2013; Kong et al. 2014, 2016; Rodrigues et al. 2014; Karker et al. 2016; Stevanovic et al. 2019).

An evaluation of the anticancer activity has revealed that the results of various in vitro treatments of cancer cells are usually accompanied by quantitative and qualitative chemical analysis to identify specialized metabolites; these are present in halophytes and are responsible for such activity (Stevanovic et al. 2019). The hexane and dichloromethane extracts of *Reaumuria vermiculata* have pointed out that they show significant anticancer activity against the A-549 lung carcinoma cells; myricetin being the main component of these investigated extracts (Karker et al. 2016).

Zygophyllum album contains triterpenes, flavonoids, and sterols. Their presence is found in extracts of different polarities, which act as anti-cancer compounds against human lung carcinoma (A-549) and colon adenocarcinoma (DLD-1) cells, in addition to their involvement as antioxidant and anti-inflammatory effects (Ksouri et al. 2013).

Rodrigues et al. (2014) have identified an active compound called "juncunol" from the *Juncus acutus* plant. They report that this compound shows selective cytotoxic activity in various human cancer cells in in vitro studies (Rodrigues et al. 2014).

A study on the *Salicornia herbacea* ethyl acetate extract has shown that phytol selectively inhibited HepG2 cells, whereas γ-linolenic acid and pentadecyl ferulate had potent antiproliferative activity against HepG2 and A549 cells (Wang et al. 2013). An anticancer potential of seed extracts on human HCT 116 and HT-29 colon cancer cells tool has been reported (Kang et al. 2011).

On the other hand, according to Kong et al. (2014), two new flavonoid glycosides are present in the *Limonium franchetii* plants. These show anticancer potential against rat C6 glioma cell lines. In another study, it has been reported that some chemicals (especially trans 3-hydroxycinnamic acid, myricetin and isorhamnetin) obtained from the extracts of *Limonium densiflorum* plants have shown strong anticancer activity against human plungous carcinoma and colon adenocarcinoma cell lines DLD-1 (Medini et al. 2015).

It has been reported that the secondary metabolite called heterocarpine reported to occur in the *Corydalis heterocarpa* halophyte taxon has significant cytotoxic effects against human gastric cancer (AGS), human colon cancer (HT-29), human fibrosarcoma (HT-1080) and human breast cancer (MCF-7) cells (Kong et al. 2016). Similarly, the methanol extract of leaves of *Imperata cylindrica* too shows anticancer activities (Hansakul et al. 2009; Bayala et al. 2014).

3.2.4 Other Bioactivities

Polyunsaturated fatty acids in halophytes have great pharmacological potential. A polyunsaturated fatty acid – linoleic acid – has been studied by many workers. Its antioxidant, anticarcinogenic, cardioprotective, anti-inflammatory and antimicrobial properties have been studied at length (Arora et al. 2017; Choi 2016). Cinnamic acid such as sinapic acid has the ability to enhance anxiolytic effects and inhibit the lipid peroxidation, therefore, it is involved in various biological activities (Yoon et al. 2007). Ferulic acid present in some halophytic grasses too is well known for its antidiabetic, anti-aging, anti-inflammatory, radioactive and anticarcinogenic properties (Mishra et al. 2014; Kumar and Pruthi 2014). The presence of cinnamic acids in halophytes confirms the significance of halophytes as food and their role in the pharmaceutical industry. Leleka et al. (2016) have studied succinic acid in halophytes as an antiradical, adaptogenic and antioxidant agent. The ursolic acid in halophytes is playing a role in reducing obesity and hepatic steatosis and increasing energy expenditure and glucose tolerance (Kunkel et al. 2012). Lupeol, a terpenoid, has been studied in vitro and in vivo by Casuga et al. (2016). They have reported its anticarcinogenic and anti-inflammatory activity (Lima et al. 2007; Saleem 2009). The antioxidant potential of halophytes has been investigated by Wong et al. (2006) using DPPH radicals inhibition capacity method. The antioxidant properties of *Salsola imbricata* have been researched by Do et al. (2014). Al-Omar et al. (2020) have investigated the antioxidant, antimicrobial, and α-amylase inhibitory activities of traditionally used halophytes. Their findings have revealed that copper, zinc, and iron in the halophytes enhance antidiabetic and antioxidant activities and *Tamarix aphylla* has anti-infection property. Under high salt concentration, the polyphenols in *Atriplex halimus* exhibit different physiological characteristics such as antioxidant, anti-allergic, anti-microbial, anti-inflammatory, anti-thrombotic, vasodilatory, and cardio-protective (Balasundram et al. 2006).

The bioactivity studies have been carried out till now on halophytes by many researchers stressing mainly on ethnopharmacological aspects. In all 78, different halophyte taxa with bioactivity potential have been identified (Table 3.2). These halophyte taxa have been determined to be effective in a total of 29 different bioactivity groups (Table 3.3). According to these findings, the halophyte taxa list includes 52 species with antioxidants, 33 species have antimicrobial, 15 are anti-inflammatory, 14 anti-cancer, and 6 species show antiviral features (Table 3.3). An evaluation made on the basis of the species reveals that *Atriplex halimus* and *Tamarix gallica* are the most bioactive among the halophytes; these are followed by *Artemisia scoparia, Cakile maritima, Limoniastrum monopetalum, Limoniun densiflorum, Chenopodium album, Mesembryanthemum edule, Nitraria retusa, Nitraria tangutorum, Spergularia marina* and *Suaeda fruticosa* (Table 3.3).

Table 3.2 List of halophyte species with potential bioactivity features

No.	Halophyte species
1	*Aegiceras corniculatum*
2	*Apocynum venetum*
3	*Artemisia campestris*
4	*Artemisia capillaris*
5	*Artemisia maritima*
6	*Artemisia santonicum*
7	*Artemisia scoparia*
8	*Arthrocnemum macrostachyum*
9	*Aster tripolium*
10	*Atriplex halimus*
11	*Atriplex hortensis*
12	*Atriplex inflata*
13	*Atriplex portulacoides*
14	*Avicennia marina*
15	*Caesalpinia crista*
16	*Cakile maritima*
17	*Camphorosma monspeliaca*
18	*Capparis decidua*
19	*Ceriops decandra*
20	*Ceriops tagal*
21	*Chenopodium album*
22	*Cladium mariscus*
23	*Corydalis heterocarpa*
24	*Crambe maritima*
25	*Cressa cretica*
26	*Crithmum maritimum*
27	*Eryngium maritimum*
28	*Excoecaria agallocha*
29	*Frankenia laevis*
30	*Frankenia pulverulenta*
31	*Glehnia littoralis*
32	*Halimione portulacoides*
33	*Inula crithmoides*
34	*Imperata cylindrica*
35	*Ipomoea pes-caprae*
36	*Juncus acutus*
37	*Limoniastrum guyonianum*
38	*Limoniastrum monopetalum*
39	*Limonium algarvense*
40	*Limonium delicatulum*

(continued)

Table 3.2 (continued)

No.	Halophyte species
41	*Limonium densiflorum*
42	*Limonium franchetii*
43	*Limonium sinense*
44	*Limonium tetragonum*
45	*Limonium vulgare*
46	*Lumnitzera littorea*
47	*Lumnitzera racemosa*
48	*Mesembryanthemum crystallinum*
49	*Mesembryanthemum edule*
50	*Nitraria retusa*
51	*Nitraria tangutorum*
52	*Plantago coronopus*
53	*Plantago lanceolata*
54	*Reaumuria vermiculata*
55	*Retama raetam*
56	*Rhizophora mucronata*
57	*Salicornia europaea*
58	*Salicornia herbacea*
59	*Salsola imbricata*
60	*Salsola kali*
61	*Salvadora persica*
62	*Sesuvium portulacastrum*
63	*Sonneratia paracaseolaris*
64	*Spergularia marina*
65	*Suaeda fruticosa*
66	*Suaeda maritima*
67	*Suaeda mollis*
68	*Suaeda monoica*
69	*Suaeda pruinosa*
70	*Suaeda salsa*
71	*Suaeda vera*
72	*Tamarix aphylla*
73	*Tamarix boveana*
74	*Tamarix gallica*
75	*Thespesia populneoides*
76	*Tribulus terrestris*
77	*Typha domingensis*
78	*Zygophyllum album*

Modified from Ksouri et al. (2012a), Slama et al. (2015), Arya et al. (2019), Parida et al. (2019), Stevanovic et al. (2019), Stanković et al. (2019), Obón et al. (2021), Stanković and Jakovljević (2021)

Table 3.3 Halophyte species with bioactivity potential

No.	Bioactivity types – Halophyte species
1	**Antioxidative activity**
1	*Artemisia capillaris*
2	*Artemisia santonicum*
3	*Artemisia scoparia*
4	*Arthrocnemum macrostachyum*
5	*Aster tripolium*
6	*Atriplex halimus*
7	*Atriplex hortensis*
8	*Atriplex inflata*
9	*Atriplex portulacoides*
10	*Cakile maritima*
11	*Camphorosma monspeliaca*
12	*Cladium mariscus*
13	*Crambe maritima*
14	*Cressa cretica*
15	*Crithmum maritimum*
16	*Frankenia laevis*
17	*Frankenia pulverulenta*
18	*Halimione portulacoides*
19	*Inula crithmoides*
20	*Ipomoea pes-caprae*
21	*Juncus acutus*
22	*Limoniastrum guyonianum*
23	*Limoniastrum monopetalum*
24	*Limonium algarvense*
25	*Limonium delicatulum*
26	*Limonium densiflorum*
27	*Limonium tetragonum*
28	*Mesembryanthemum crystallinum*
29	*Mesembryanthemum edule*
30	*Nitraria tangutorum*
31	*Plantago coronopus*
32	*Plantago lanceolata*
33	*Retama raetam*
34	*Salicornia europaea*
35	*Salicornia herbacea*
36	*Salsola imbricata*
37	*Salsola kali*
38	*Salvadora persica*
39	*Sesuvium portulacastrum*

(continued)

Table 3.3 (continued)

No.	Bioactivity types – Halophyte species
40	*Spergularia marina*
41	*Suaeda fruticosa*
42	*Suaeda maritima*
43	*Suaeda mollis*
44	*Suaeda pruinosa*
45	*Suaeda salsa*
46	*Suaeda vera*
47	*Tamarix aphylla*
48	*Tamarix boveana*
49	*Tamarix gallica*
50	*Thespesia populneoides*
51	*Typha domingensis*
52	*Zygophyllum album*
2	**Antimicrobial activity**
1	*Aegiceras corniculatum*
2	*Apocynum venetum*
3	*Artemisia campestris*
4	*Artemisia maritima*
5	*Artemisia scoparia*
6	*Artemisia santonicum*
7	*Atriplex halimus*
8	*Avicennia marina*
9	*Cakile maritima*
10	*Ceriops decandra*
11	*Ceriops tagal*
12	*Crithmum maritimum*
13	*Eryngium maritimum*
14	*Excoecaria agallocha*
15	*Frankenia laevis*
16	*Limoniastrum guyonianum*
17	*Limoniastrum monopetalum*
18	*Limonium delicatulum*
19	*Limonium densiflorum*
20	*Limonium vulgare*
21	*Lumnitzera littorea*
22	*Lumnitzera racemosa*
23	*Mesembryanthemum edule*
24	*Retama raetam*
25	*Salsola kali*
26	*Sesuvium portulacastrum*
27	*Spergularia marina*
28	*Suaeda fruticosa*

(continued)

Table 3.3 (continued)

No.	Bioactivity types – Halophyte species
29	*Suaeda monoica*
30	*Suaeda pruinosa*
31	*Suaeda salsa*
32	*Tamarix boveana*
33	*Tamarix gallica*
3	**Anticancerogenic and/or anti-tumor activity**
1	*Artemisia scoparia*
2	*Cakile maritima*
3	*Corydalis heterocarpa*
4	*Juncus acutus*
5	*Imperata cylindrica*
6	*Limoniastrum monopetalum*
7	*Limonium densiflorum*
8	*Limonium franchetii*
9	*Reaumuria vermiculata*
10	*Salicornia herbacea*
11	*Sonneratia paracaseolaris*
12	*Suaeda fruticosa*
13	*Tamarix gallica*
14	*Zygophyllum album*
4	**Antiviral activity**
1	*Avicennia marina*
2	*Limonium densiflorum*
3	*Limonium sinense*
4	*Limonium tetragonum*
5	*Rhizophora mucronata*
6	*Sonneratia paracaseolaris*
5	**Anti-inflammatory activity**
1	*Artemisia capillaris*
2	*Artemisia scoparia*
3	*Atriplex halimus*
4	*Chenopodium album*
5	*Eryngium maritimum*
6	*Glehnia littoralis*
7	*Limoniastrum monopetalum*
8	*Limonium densiflorum*
9	*Nitraria retusa*
10	*Salicornia herbacea*
11	*Spergularia marina*
12	*Suaeda fruticosa*
13	*Plantago lanceolata*
14	*Tamarix gallica*

(continued)

Table 3.3 (continued)

No.	Bioactivity types – Halophyte species
15	*Zygophyllum album*
6	**Anti-adipogenic activity**
1	*Spergularia marina*
7	**Anti-allergenic activity**
1	*Atriplex halimus*
8	**Antiarrhythmic agent**
1	*Nitraria retusa*
2	*Nitraria tangutorum*
9	**Antidepressant agent**
1	*Artemisia scoparia*
10	**Anti-diabetic activity**
1	*Capparis decidua*
2	*Tamarix aphylla*
11	**Antidiarrheic activity**
1	*Tamarix gallica*
12	**Anti-dysentric activity**
1	*Limoniastrum monopetalum*
13	**Antifungal activity**
1	*Cakile maritima*
2	*Mesembryanthemum edule*
14	**Antihyperlipidaemic activity**
1	*Capparis decidua*
15	**Anti-hypertensive activity**
1	*Apocynum venetum*
2	*Tribulus terrestris*
16	**Anti-hypertonia activity**
1	*Salsola kali*
17	**Antimalarial agent**
1	*Caesalpinia crista*
18	**Antineuropathic activity**
1	*Nitraria retusa*
2	*Nitraria tangutorum*
19	**Antinociceptive activity**
1	*Chenopodium album*
2	*Eryngium maritimum*
20	**Antiparasitic properties**
1	*Atriplex halimus*
2	*Chenopodium album*
21	**Anti-proliferative activity**
1	*Glehnia littoralis*
22	**Antipruritic activity**
1	*Chenopodium album*

(continued)

Table 3.3 (continued)

No.	Bioactivity types – Halophyte species
23	**Antipyretic agent**
1	*Artemisia capillaris*
2	*Glehnia littoralis*
24	**Anti-scorbutic activity**
1	*Cakile maritima*
2	*Crithmum maritimum*
25	**Antiseptic agents**
1	*Mesembryanthemum crystallinum*
2	*Mesembryanthemum edule*
3	*Tamarix gallica*
26	**Antispasmodic agent**
1	*Nitraria retusa*
2	*Nitraria tangutorum*
27	**Anti-thrombotic activity**
1	*Atriplex halimus*
28	**Antiulcer activity**
1	*Excoecaria agallocha*
29	**Anti-cholinesterase activity**
1	*Arthrocnemum macrostachyum*
2	*Crithmum maritimum*
3	*Halimione portulacoides*
4	*Salicornia europaea*

Modified from Ksouri et al. (2012a), Slama et al. (2015), Arya et al. (2019), Parida et al. (2019), Stevanovic et al. (2019), Stanković et al. (2019), Obón et al. (2021), Saleh et al. (2021), Stanković and Jakovljević (2021)

References

Abdel-Aziz SM, Mouafi FE, Moustafa YA, Abdelwahed NAM (2016) Medicinal importance of mangrove plants. In: Garg N, Abdel-Aziz SM, Aeron A (eds) Microbes in food and health. Springer, Cham, Switzerland, pp 77–96

Ahmad P, Ozturk M, Sharma S, Gucel S (2013) Effect of sodium carbonate induced salinity- alkalinity on some key osmoprotectants protein profile antioxidant enzymes and lipid peroxidation in two mulberry (*Morus alba* L.) cultivars. J Plant Interact 9(1):460–467

Al-Omar MS, Mohammed HA, Mohammed SA et al (2020) Anti-microbial, anti-oxidant, and α-amylase inhibitory activity of traditionally-used medicinal herbs: a comparative analyses of pharmacology, and phytoconstituents of regional halophytic plants' diaspora. Molecules 25(22):5457

Alonso-Amelot EM, Oliveros A, Calcagno-Pisarelli PM (2004) Phenolics and condensed tannins in relation to altitude in neotropical Pteridium spp. A field study in the Venezuelan Andes. Biochem Syst Ecol 32:969–981

Arora S, Kumar G, Meena S (2017) Gas chromatography-mass spectroscopy analysis of root of an economically important plant, Cenchrus ciliaris L. from Thar desert, Rajasthan (India). Asian J Pharm Clin Res 10:64–69

Arya SS, Devi S, Ram K et al (2019) Halophytes: the plants of therapeutic medicine. In: Hasanuzzaman M et al (eds) Ecophysiology, abiotic stress responses and utilization of halophytes. Springer, Singapore, pp 271–287

Balasundram N, Sundram K, Samman S (2006) Phenolic compounds in plants and agri ındustrial by-products: antioxidant activity, occurrence, and potential uses. Food Chem 99:191–203

Bayala B, Bassole IH, Scifo R et al (2014) Anticancer activity of essential oils and their chemical components-a review. Am J Cancer Res 4:591–607

Bernal J, Mendiola JA, Ibánez E, Cifuentes A (2011) Advanced analysis of nutraceuticals. J Pharm Biomed Anal 55:758–774

Bernstein N, Shoresh M, Xu Y, Huang B (2010) Involvement of the plant antioxidative response in the differential growth sensitivity to salinity of leaves vs roots during cell development. Free Radic Biol Med 49:1161–1171

Bose J, Rodrigo-Moreno A, Shabala S (2014) ROS homeostasis in halophytes in the context of salinity stress tolerance. J Exp Bot 65:1241–1257

Boughalleb F, Denden M (2011) Physiological and biochemical changes of two halophytes, Nitraria retusa (Forssk.) and Atriplex halimus L. under increasing salinity. Agric J 6:327–339

Buhmann A, Papenbrock J (2013) Biofiltering of aquaculture effluents by halophytic plants: basic principles, current uses and future perspectives. Environ Exp Bot 92:122–133

Cagliari A, Margis R, Dos SMF et al (2011) Biosynthesis of triacylglycerols (TAGs) in plants and alga. Int J Plant Biol 2:40–52

Caparrós PG, Ozturk M, Gul A et al (2022) Halophytes have potential as heavy metal phytoremediators: A comprehensive review. Environ Exp Bot 193:104666

Casuga FP, Castillo AL, Corpuz MJAT (2016) GC-MS analysis of bioactive compounds present in different extracts of an endemic plant Broussonetia luzonica (Blanco) (Moraceae) leaves. Asian Pac J Trop Biomed 6:957–961

Chen THH, Murata N (2011) Glycinebetaine protects plants against abiotic stress: mechanisms and biotechnological application. Plant Cell Environ 34:1–20

Choi WH (2016) Evaluation of anti-tubercular activity of linolenic acid and conjugated-linoleic acid as effective inhibitors against mycobacterium tuberculosis. Asian Pac J Trop Med 9:125–129

Cowan MM (1999) Plant products as antimicrobial agents. Clin Microbiol Rev 12:564–582

Crozier A, Clifford NM, Ashihara H (2006) Plant secondary metabolites: occurrence, structure, and role in the human diet. Blackwell Publishing, Oxford

Cushnie TPT, Lamb AJ (2005) Antimicrobial activity of flavonoids. Int J Antimicrob Agents 26:343–356

Dajic Stevanovic Z, Janackovic P, Stankovic M (2014) Are there still neglected medicinal plants beyond official and traditional consideration? In: Dajic Stevanovic Z, Ibraliu A (eds) Proceedings of the 8th conference on medicinal and aromatic plants of Southeast European Countries (8th CMAPSEEC), 19-22 May 2014. Albanian Academy of Science, Agricultural University of Tirana and AMAPSEEC, Durrës, Albania, p 13–23

de Jesus Cortes-Sanchez A, Hernández-Sánchez H, Jaramillo-Flores ME (2013) Biological activity of glycolipids produced by microorganisms: new trends and possible therapeutic alternatives. Microbiol Res 168(1):22–32

Do QD, Angkawijaya AE, Tran-Nguyen PL et al (2014) Effect of extraction solvent on total phenol content, total flavonoid content, and antioxidant activity of Limnophila aromatica. J Food Drug Anal 22:296–302

Duangmano S, Dakeng S, Jiratchariyakul W et al (2010) Antiproliferative effects of cucurbitacin B in breast cancer cells: down-regulation of the c-Myc/hTERT/Telomerase pathway and obstruction of the cell cycle. Int J Mol Sci 11:5323–5338

Falleh H, Ksouri R, Medini F et al (2011) Antioxidant activity and phenolic composition of the medicinal and edible halophyte *Mesembryanthemum edule* L. Ind Crop Prod 34:1066–1071

Fan TWM, Colmer TD, Lane AN, Higashi RM (1993) Determination of metabolites by H NMR and GC: analysis for organic osmolytes in crude tissue extracts. Anal Biochem 214:260–271

Flowers TJ, Colmer TD (2008) Salinity tolerance in halophytes. New Phytol 179:945–963

Galvani A (2007) The challenge of the food sufficiency through salt tolerant crops. Rev Environ Sci Biotechnol 6:3–16

Garcia-Caparros P, De Filippis L, Gul A et al (2021) Oxidative stress and antioxidant metabolism under adverse environmental conditions: a review. Bot Rev 87(4):421–466

Gourine N, Bombarda MI, Nadjemi B et al (2010) Antioxidant activities and chemical composition of essential oil of *Pistacia atlantica* from Algeria. Ind Crop Prod 31:203–208

Hameed M, Ashraf M, Ahmad MSA, Naz N (2010) Structural and functional adaptations in plants for salinity tolerance. In: Asharaf M, Ozturk M, Ahmad MSA (eds) Plant adaptation and phytoremediation. Springer, Berlin Germany, pp 151–173

Hansakul P, Wongnoppavich A, Ingkaninan K et al (2009) Apoptotic induction activity of Dactyloctenium aegyptium (L.) P.B. and Eleusine indica (L.) Gaerth. extracts on human lung and cervical cancer cell lines. Songklanakarin J Sci Technol 31:273–279

Jallali I, Megdiche W, M'Hamdi B et al (2012) Changes in phenolic composition and antioxidant activities of the edible halophyte Crithmum maritimum L. with physiological stage and extraction method. Acta Physiol Plant 34:1451–1459

Jayakannan M, Bose J, Baboutina O et al (2013) Salicyclic acid improves salinity tolerance in Arabidopsis by restoring membrane potential and preventing salt-induced K+ loss via a GORK channel. J Exp Bot 64:2255–2268

Jdey A, Falleh H, Jannet SB et al (2017) Phytochemical investigation and antioxidant, antibacterial and anti-tyrosinase performances of six medicinal halophytes. S Afr J Bot 112:508–514

Kang S, Kim D, Lee BH et al (2011) Antioxidant properties and cytotoxic effects of fractions from glasswort (*Salicornia herbacea*) seed extracts on human intestinal cells. Food Sci Biotechnol 20:115–122

Karker M, Falleh H, Msaada K et al (2016) Antioxidant, anti-inflammatory and anticancer activities of the medicinal halophyte *Reaumuria vermiculata*. Clin. Exp. Med Sci J 15:297–307

Khan DA, Hamdani SDA, Iftikhar S et al (2021) Pharmacoinformatics approaches in the discovery of drug-like antimicrobials of plant origin. J Biomol Struct Dyn. https://doi.org/10.1080/07391102.2021.1894982

Khoo BY, Chua SL, Balaram P (2010) Apoptotic effects of chrysin in human cancer cell lines. Int J Mol Sci 11:2188–2199

Kong NN, Fang ST, Wang JH et al (2014) Two new flavonoid glycosides from the halophyte *Limonium franchetii*. J Asian Nat Prod Res 16(4):370–375

Kong CS, Kim YA, Kim H, Seo Y (2016) Evaluation of a furochromone from the halophyte *Corydalis heterocarpa* for cytotoxic activity against human gastric cancer (AGS) cells. Food Funct 7(12):4823–4829

Ksouri R, Falleh H, Megdiche W et al (2009) Antioxidant and antimicrobial activities of the edible medicinal halophyte *Tamarix gallica* L. and related polyphenolic constituents. Food Chem Toxicol 47:2083–2091

Ksouri R, Ksouri WM, Jallali I et al (2012a) Medicinal halophytes: potent source of health promoting biomolecules with medical, nutraceutical and food applications. Crit Rev Biotechnol 32:289–326

Ksouri R, Smaoui A, Isoda H, Abdelly C (2012b) Utilization of halophyte species as new sources of bioactive substances. J Arid Land Stud 22:41–44

Ksouri WM, Medini F, Mkadmini K et al (2013) LC-ESI-TOF-MS identification of bioactive secondary metabolites involved in the antioxidant, anti-inflammatory and anticancer activities of the edible halophyte *Zygophyllum album* Desf. Food Chem 139(1-4):1073–1080

Kumar N, Pruthi V (2014) Potential applications of ferulic acid from natural sources. Biotechnol Rep 4:86–93

Kunkel SD, Elmore CJ, Bongers KS et al (2012) Ursolic acid increases skeletal muscle and brown fat and decreases diet-induced obesity, glucose intolerance and fatty liver sisease. PLoS One 7:e39332

Lee J, Kong CS, Jung M et al (2011) Antioxidant activity of the halophyte *Limonium tetragonum* and its major active components. Biotechnol Bioprocess Eng 16:992–999

Leleka M, Zalis'ka O, Kozyr G (2016) Screening research of pharmaceutical compositions based on succinic acid, ascorbic acid and rutin. J Pharm Pharmacol 4:486–491

Lima LM, Perazzo FF, Tavares CJC, Bastos JK (2007) Anti-inflammatory and analgesic activities of the ethanolic extracts from Zanthoxylum riedelianum (Rutaceae) leaves and stem bark. J Pharm Pharmacol 59:1151–1158

Lou Z, Wang H, Zhu S et al (2011) Antibacterial activity and mechanism of action of chlorogenic acid. J Food Sci 76(6):398–403

Mansour MMF, Ali EF (2017) Evaluation of proline fuctions in saline conditions. Phytochemistry 140:52–68

Maria VF, Maria AFF, Diana CGP (2019) Halophytic grasses, a new source of nutraceuticals? A review on their secondary metabolites and biological activities. Int J Mol Sci 20(5):1067

Medini F, Bourgou S, Lalancette GK et al (2015) Phytochemical analysis, antioxidant, anti-inflammatory, and anticancer activities of the halophyte *Limonium densiflorum* extracts on human cell lines and murine macrophages. S Afr J Bot 99:158–164

Menzel U, Lieth H (1999) Halophyte database vers. 2.0. Halophyte uses in different climates I: Ecological and ecophysiological studies. In: Lieth H et al (eds) Progress in biometeorology. Backhuys Publishers, Leiden, pp 77–88

Meot-Duros L, Le Floch G, Magne C (2008) Radical scavenging, antioxidant and antimicrobial activities of halophytic species. J Ethnopharmacol 116:258–262

Minocha R, Majumdar R, Minocha SC (2014) Polyamines and abiotic stress in plants: a complex relationship. Front Plant Sci 5:1–17

Mishra K, Ojha KS et al (2014) Protective effect of ferulic acid on ionizing radiation induced damage in bovine serum albumin. Int J Radiat Res 12:113–121

Mohammed HA (2020) The valuable ımpacts of halophytic genus Suaeda; nutritional, chemical, and biological values. Med Chem 16:1

Mostafa MG, Hossain MA, Fujita M (2015) Trehalose pretreatment induces salt tolerance in rice (Oryza sativa L.) seedlings: oxidative damage and co-induction of antioxidant defense and glyoxalase systems. Protoplasma 252:461–475

Obón C, Rivera D, Verde A, Alcaraz F (2021) Ethnopharmacology and medicinal uses of extreme halophytes. In: Grigore M-N (ed) Handbook of halophytes: from molecules to ecosystems towards biosaline agriculture. Springer Nature, Switzerland, pp 2707–2735

Oueslati S, Ksouri R, Falleh H et al (2012a) Phenolic content, antioxidant, anti-inflammatory and anticancer activities of the edible halophyte *Suaeda fruticosa* Forssk. Food Chem 132(2):943–947

Oueslati S, Trabelsi N, Boulaaba M et al (2012b) Evaluation of antioxidant activities of the edible and medicinal *Suaeda* species and related phenolic compounds. Ind Crop Prod 36(1):513–518

Ozturk M, Szaniawiski RK (1981) Root temperature stress and proline content in leaves and roots of two ecologically different plant species. Plant Physiol 102:375–377

Ozturk M, Sato T, Takahashi N (1986) Proline accumulation in shoots and roots of some eco-physiologically different plants under root temperature stress. Environ Control Biol (Japan) 24:79–85

Ozturk M, Turkyilmaz B, Gucel S, Guvensen A (2011) Proline accumulation in some coastal zone plants of the aegean region of Turkey. Eur J Plant Sci Biotechnol 5(Special Issue 2):54–56

Ozturk M, Gucel S, Altay V et al (2019) Clustering of halophytic species from Cyprus based on ionic contents. Pyton Int J Exp Bot 88(3):223–238

Pandey KB, Rizvi SI (2009) Plant polyphenols as dietary antioxidants in human health and disease. Oxidative Med Cell Longev 2:270–278

Parida AK, Kumari A, Rangani J, Patel M (2019) Halophytes: potential resources of coastal ecosystems and their economic, ecological and bioprospecting significance. In: Hasanuzzaman M et al (eds) Halophytes and climate change: adaptive mechanisms and potential uses. CAB International, pp 287–323

Patel MK, Mishra A, Jha B (2016) Untargeted metabolomics of halophytes. In: Kim S (ed) Marine Omics: principles and applications. CRC Press, Boca Raton, FL, pp 309–325

Plouguerné E, da Gama BAP, Pereira RC, Barreto-Bergter E (2014) Glycolipids from seaweeds and their potential biotechnological applications. Front Cell Infect Microbiol 4. https://doi.org/10.3389/fcimb.2014.00174

Povichit N, Phrutivorapongkul A, Suttajit M et al (2010) Phenolic content and in vitro inhibitory effects on oxidation and protein glycation of some Thai medicinal plant. Pak J Pharm Sci 23(4):403–408

Qasim M, Gulzar S, Khan MA (2011) Halophytes as medicinal plants. In: Öztürk M, Mermut AR, Celik A (eds) Urbanisation, land use, land degradation and environment. Daya Publishing House, New Delhi, pp 330–343

Qian D, Zhao Y, Yang G, Huang L (2017) Systematic review of chemical constituents in the genus *Lycium* (Solanaceae). Molecules 22:1–33

Quideau S, Deffieux D, Douat-Casassus C, Pouysegu L (2011) Plant polyphenols: chemical properties, biological activities, and synthesis. Angew Chem Int Ed Eng 50:586–621

Raza MM, Ullah S, Aziz T et al (2019) Alleviation of salinity stress in maize using silicon nutrition. Not Bot Horti Agro Botanica 47(4):1340–1347

Rejeb KB, Abdelly C, Savoure A (2014) How reactive oxygen species and proline face stress together. Plant Physiol Biochem 80:278–284

Ren W, Qiao Z, Wang H et al (2003) Flavonoids: promising anticancer agents. Med Res Rev 23(4):519–534

Rodrigues MJ, Gangadhar KN, Vizetto-Duarte C et al (2014) Maritime halophyte species from southern Portugal as sources of bioactive molecules. Marine Drugs 12(4):2228–2244

Rodrigues MJ, Soszynski A, Martins A et al (2015) Unravelling the antioxidant potential and the phenolic composition of different anatomical organs of the marine halophyte *Limonium algarvense*. Ind Crop Prod 77:315–322

Sakamoto A, Murata N (2000) Genetic engineering of glycinebrtaine synthesis in plants: current status and implications for enhancement of stress tolerance. J Exp Bot 51:81–88

Saleem M (2009) Lupeol, a novel anti-inflammatory and anti-cancer dietary triterpene. Cancer Lett 285:109–115

Saleh IA, Usman K, Abu-Dieyeh MH (2021) Halophytes as ımportant sources of antioxidants and anti-cholinesterase compounds. In: Grigore M-N (ed) Handbook of halophytes: from molecules to ecosystems towards biosaline agriculture. Springer Nature, Switzerland, pp 1931–1951

Scheiber IF, Mercer JF, Dringen R (2014) Metabolism and functions of copper in brain. Prog Neurobiol 116:33–57

Shabala S, Mackay A (2011) Ion transport in halophytes. Adv Bot Res 57:151–199

Shahrasbi S, Pirasteh-Anosheh H, Emam Y et al (2021) Elucidating some physiological mechanisms of salt tolerance in *Brassica napus* L. seedlings induced by seed priming with plant growth regulators. Pak J Bot 53(2):34

Slama I, Abdelly C, Bouchereau A et al (2015) Diversity, distribution and roles of osmoprotective compounds accumulated in halophytes under abiotic stress. Ann Bot 115:433–447

Sokolowska-Krzaczek A, Skalicka-Wozniak K, Czubkowska K (2009) Variation of phenolic acids from herb and roots of *Salsola kali* L. Acta Soc Bot Pol 78:197–201

Stanković M, Jakovljević D (2021) Phytochemical diversity of halophytes. In: Grigore M-N (ed) Handbook of halophytes: from molecules to ecosystems towards biosaline agriculture. Springer Nature, Switzerland, pp 2089–2114

Stanković M, Jakovljević D, Stojadinov M, Stevanović ZD (2019) Halophyte species as a source of secondary metabolites with antioxidant activity. In: Hasanuzzaman M et al (eds) Ecophysiology, abiotic stress responses and utilization of halophytes. Springer, Singapore, pp 289–312

Stevanovic Z, Stankovic MS, Stankovic J et al (2019) Use of halophytes as medicinal plants: phytochemical diversity and biological activity. In: Hasanuzzaman M, Shabala S, Fujita M (eds) Halophytes and climate change: adaptive mechanisms and potential uses. CAB International, pp 343–358

Stöckigt J, Sheludko Y, Unger M et al (2002) High-performance liquid chromatographic, capillary electrophoretic and capillary electrophoretic-electrospray ionisation mass spectrometric analysis of selected alkaloid groups. J Chromatogr A 967:85–113

Stuchlík M, Žák S (2002) Vegetable lipids as components of functional foods. Biomedical Papers 146:3–10

Trabelsi N, Oueslati S, Falleh H et al (2012) Isolation of powerful antioxidants from the medicinal halophyte *Limoniastrum guyonianum*. Food Chem 135(3):1419–1424

Turkyilmaz Unal B, Guvensen A, Esiz Dereboylu A, Ozturk M (2013) Variations in the proline and total protein contents in *Origanum sipyleum* L. from different altitudes of Spil Mountain, Turkey. Pak J Bot 45(S1):571–576

Wang H, Wang X, You J et al (2007) Comparative analysis of allantoin, quercetin, and 1-methyl-1,2,3,4-tetrahydro-β-carboline-3-carboxylic acid in *Nitraria tangutorum* Bobr. seed by HPLC-APCI-MS and CE. J Liq Chromatogr Relat Technol 30:363–376

Wang X, Zhang M, Zhao Y et al (2013) Pentadecyl ferulate, a potent antioxidant and antiproliferative agent from the halophyte *Salicornia herbacea*. Food Chem 141(3):2066–2074

Widhalm JR, Dudareva N (2015) A familiar ring to it: biosynthesis of plant benzoic acids. Mol Plant 8:83–97

Wink M (2003) Evolution of secondary metabolites from an ecological and molecular phylogenetic perspective. Phytochemistry 64:3–19

Wong SP, Leong LP, Koh JHW (2006) Antioxidant activities of aqueous extracts of selected plants. Food Chem 99:775–783

Xiong L, Schumaker KS, Zhu JK (2002) Cell signaling during cold, drought and salt stress. Plant Cell 2002(Supplement):165–183

Xu DP, Li Y, Meng X et al (2017) Natural antioxidants in foods and medicinal plants: extraction, assessment and resources. Int J Mol Sci 18(1):96

Yoon BH, Jung JW, Lee JJ et al (2007) Anxiolytic-like effects of sinapic acid in mice. Life Sci 81:234–240

Younessi-Mehdikhanlou M, Ozturk M, Jafarpour S, Mahna N (2022) Exploitation of next generation sequencing technologies for unraveling metabolic pathways in medicinal plants: a concise review. Ind Crop Prod 178(114669):1–16

Chapter 4
Ethnic Aspects of Halophytes and Importance in the Economy

4.1 Ethnic Aspects

Halophytes include plant species that are adapted to live especially in saline soils. About 955 million hectors of land are salt-affected worldwide and the Southwestern Asian deserts constitute the most part of this land. Nearly 30% of the semi-arid irrigated areas of Asia are facing salinization. Afghanistan, Bahrain, Iran, Iraq, Jordan, Kuwait, Lebanon, Oman, Pakistan, Palestine, Qatar, Saudi Arabia, and Yemen are mostly arid countries having large sandy and gravel deserts. Some parts of Asia are below sea level due to continental rifting, like the Dead Sea which lies between Jordan and Israel. Anatolian Peninsula is located between the Mediterranean and the Black Sea, while Arabian Peninsula is located between the Persian Gulf and the Red Sea. Gravel and sandy plains cover the Arabian Peninsula in the Southwest. Rubal Khali is also located here, which is included among the world's large sand deserts. Iran's central plateau has large saline areas. Zagros mountainous range in Iran is an arid area, divided into the salt desert (Dasht-e-Kavir) and Dasht-e-Loot. Umma's Samim is the large salt inland plain located in the western part of Oman. Most of the plain areas in Asia receive less than 400 mm of precipitation in a year. Some regions in Southwest Asia have insufficient resources of water. The Euphrates and Tigris rivers flow through Syria, Iraq, and Turkiye and are considered critical for the agriculture in the area. Most Asian deserts have large saline areas with salt flats. The information on the accurate identification of salt-tolerant plants and their potential medicinal and non-medicinal utilization and other economic benefits is very weak and scattered in most regions of Asia. This information is necessary for the research on rangeland management, biosaline agriculture, coastal management and saline, and arid habitat restoration.

M. Ozturk et al., *Halophyte Plant Diversity and Public Health*,
https://doi.org/10.1007/978-3-031-21944-3_4

Saline habitats are regarded as one of the most fragile ecosystems. Many floristic, ecological, and ecophysiological studies have been carried out on these habitats by different workers. In addition to these studies, lately some initiative activities have been started emphasizing their sustainable use, management as well as importance of these sensitive habitats. In accordance with this aim agriculture, agroforestry and eco-tourism have started to come to the forefront. Although these habitats seem harsh for alternative agriculture, the possible uses and potential evaluations of halophytes, the characteristic natural plants of these habitats, allow them to provide different ecosystem services (Yensen 2006; Luković et al. 2021). Among these ecosystem services, supply, regulatory, and cultural services top the list. In supply services, halophytes are considered as an alternative potential for genetic resources of medicinal plants, food and/or vegetables, animal feed, biofuel and fuelwood, lumber, fiber, and for ornamental flowers. Regulatory services refer to the significant benefits halophytes and saline habitats provide such as climate regulation, coastal reclamation and protection, land reclamation and landscaping, carbon sequestration, biodiversity preservation, honey production, bioremediation or phytoremediation, pollination and pest control. Cultural services, on the other hand, are based on the evaluation of these ecosystems in terms of alternative potential areas these are of recreational value and have aesthetic sides as well that they provide to human beings (Dajić Stevanović et al. 2019; Luković et al. 2021; Radulovich and Umanzor 2021).

4.2 Economic Significance

Halophytes are of great importance for restoration of saline and degraded lands. They have a potential for biomass production for renewable energy, as cash crops, for medicinal uses, as fodder, for CO_2 sequestration, and in building materials. In arid and semi-arid zones, mostly vegetation is sparse due to extreme and unpredictable temperature ranges but the potential of medicinal plants in these areas has an economic and social potential. Most of these areas are unexploited because of a lack of water for agriculture. The economic potential of saline vegetation is greatly imperative as it reduces the loss caused by the water and soil salinization.

In view of their ecological abilities, halophytes thrive well on saline soils. These plants are also very important economically (Aronsen 1989; Ahmad et al. 2021). Some halophytes are consumed as food, used as medicine and feed, which reveals their great potential (Aronsen 1989; Ahmad et al. 2021). However, these natural resources are decreasing gradually, and some representative species have reached to the point of extinction, the reason being population pollution which have tremendous anthropogenic impacts. The halophytes have been neglected for a long time as natural resources (Ahmad et al. 2021). Their utilization as food, fodder and fuelwood have been reported in most regions of Asia (Rozema and Flowers 2008; Ozturk et al. 2014, 2016, 2019a). Their use as medicinal plants has been acknowledged in many research papers published in various parts of Asia together with their healing potential (Ahmad et al. 2014; Nadembega et al. 2011; Malik et al. 2021; Ozturk et al. 2021a, 2022). Ghazanfar (1995) in his book on Middle East medicinal plants has reported different beneficial uses of halophytes. She has mentioned that *Phragmites* and *Typha* species are useful for making boats, mats and thatching, sa well as the medicinal uses of halophytes by native people as a support towards their regional socioeconomic conditions. Halophytes can be cultivated as energy crops, fodder, and food crops on unused salinized lands, thereby resting them for productive purposes by planting halophytes (Boyko and Boyko 1964; Ozturk et al. 2014, 2016).

4.2.1 Halophytes as a Source of Medicinal Plants

Ethnobotanical assessments of medicinal plants provide knowledge about their traditional uses, conservation, indigenous communities, and the necessity for the herbal drug development (Heinrich and Gibbons 2001; Vitalini et al. 2013; Altay et al. 2015a, b; Malik et al. 2021; Ozturk et al. 2021a, 2022). These surveys are crucial to enlighten the important plant species, mainly for developing new crude drugs (Cox 2000; Leonti et al. 2002; Ozturk et al. 2017a; Pleskanovskaya et al. 2019; Masoodi et al. 2020). Indigenous knowledge has been held by the traditional societies as a fundamental to their health and social interests (Cotton 1996; Getahun 1976; Malik et al. 2021). Moreover, the ethnomedicinal investigations support the communities in sociocultural and socioeconomic contexts, being important for the conservation of global plant heritages (Sanz-Biset et al. 2009; Younessi-Hamzekhanlu et al. 2020; Malik et al. 2021; Ozturk et al. 2021a, 2022).

Several plants with great therapeutic potential have been broadly exploited in different regions of the world. Ethnic knowledge about medicinal plants confirms their therapeutic activities against many diseases (Balunas and Kinghorn 2005; Nadembega et al. 2011; Malik et al. 2021). Several ailments have been cured by plant-based medications in arid regions (Yaseen et al. 2015). Most people in rural areas in the developing world do not have access to drugs, therefore, they use traditional herbal medicines to cure diseases (Calixto 2005; Malik et al. 2021; Ozturk et al. 2021a, 2022). Almost 4 billion humans in developing countries are consistently exploiting the therapeutic potential of plant species. The medicinally valuable plants from the wild have been utilized since ancient times by different ethnic groups in different areas of the world (Ahmad et al. 2020; Malik et al. 2021; Ozturk et al. 2021a, 2022). Several pharmacological and ethnobotanical studies have been carried out on the preference and utilization of plants by several ethnic groups globally (Jeruto et al. 2008; Walter et al. 2011; Altay and Çelik 2011; Altay and Karahan 2012; Ozturk et al. 2012; Altay et al. 2015a; Altay and Karahan 2017; Imanberdieva et al. 2018a, b; Ozturk et al. 2017b, 2017c, 2018a, b, c, d, e, f, g; Letchamo et al. 2018; Malik et al. 2021; Ozturk et al. 2021a, 2022).

The medicinal plants are still recognized as primarily health care in underdeveloped communities because of their cultural preferences while in developed countries these give an alternative for health care (Svarstad and Dhillion 2000; Malik et al. 2021). The report shows that 70–80% of the people depend on the ethnobotanical and ethno-medicinal uses of medicinal plants in the world (Muthu et al. 2006). A rough estimate is that 35,000–75,000 medicinal plants provide the basis for health care systems worldwide (Farnsworth and Soejarto 1991). Herbal pharmaceuticals play significant role in the cure of number of human ailments (Rehecho et al. 2011). The ethnobotanical studies have gained significant attention among the scientific communities concerning medicinal plants utilization (Tripathi et al. 2017). A number of ethnomedicinal surveys have been converted into conservation and health care programs (Quamar and Bera 2014). The ways of plant utilization are diverse in various communities around the globe (Abebe et al. 2003; Kassaye et al. 2006; Malik et al. 2021; Ozturk et al. 2021a, 2022).

Many drugs available on the market are synthetic analogs of isolated compounds from plants (Amjad et al. 2017). Traditional phytotherapy is still popularly practiced in many rural communities and handed down from one generation to the next in spite of the variety of clinical agents developed by the pharmaceutical industries (Abbasi et al. 2010). The extensive cultural value studies and their systematic documentation have been placed on high priority to highlight the pharmaceutical potential of plants and to explore their new properties during the past decades. The traditional studies on the plants plays important role in farming, nutraceutical and pharmaceutical industries (Cox 2000; Idrisi et al. 2010). The interactions of the number of chemical constituents in medicinal plants cause harmony and synergy in the human body and heal the body tissues progressively. The drug industries have developed medicines using isolated chemical compounds from medicinal plants which have physiological responses (Shinwari and Gilani 2003). To ascertain the therapeutic uses of plant-based drugs, it is necessary to conduct the pharmacological and pharmacognostic studies and explore their uses (Ahmad et al. 2009).

The increasing research interest in halophytes reflects their importance as valuable resources as a cash crop (Khan et al. 2009). The medicinal uses of halophytes have been recommended for the increasing population since times (Rozema and Flowers 2008). An increasing population in Asia is facing the challenges of food security and increase in growth is exacerbated by fluctuations in climate change. The high temperature, variable precipitation rates, and increase in salinity are major impacts of climate change (Flowers and Muscolo 2015). The fresh water and land of good quality are always used for the cultivation of crops and food but the real potential lies in the reclamation saline lands for cultivation of cash crop halophytes. Out of 6.5 million hectares of salt-affected land, two million hectares are moderately affected, which can be evaluated for growing "halophytes" as these areas present diverse environmental conditions and halophytes growing on such habitats points out to their ability to tolerate different salts (Song et al. 2008).

Most of the saline areas in Asia are underdeveloped and the rural communities depend on native flora and traditional therapeutics for health care. The ethnomedicinal significance of halophytes stresses the fact that the halophytes have potential as a valuable resource (Khan et al. 2009). These plants give many alternatives in various countries (Khan and Duke 2001). According to Koyro et al. (2011), these are of great ecological, industrial and agricultural value. Indigenous knowledge is decreasing day by day due to rapid land degradation, modern cultural exposure, urban development, and war effects much the rural communities, the traditional knowledge about medicinal plants survives due to its transformation from one generation to the next generation (Figueiredo et al. 1993; Kayani et al. 2015; Manandhar 1995; Tabuti et al. 2003). The domestication of salt-tolerant plant species is of great significance. The salinity has not threatened the food availability for humans. The continuous increase in population and climate change at the global level are predicting a decrease of resources in the next century which will push us towards the need for new avenues for food availability and health security.

Asian countries have a variety of climate zones and topographic regions and are endowed with a diversity of halophytic medicinal plants. The parochial knowledge and herbal utilization reflect the different natural resources present in different countries. Ahmad and Waseem (2006) have reported that older people utilize more the traditional knowledge about medicinal plant species than younger people. The cultivation and conservation of such natural resources can uplift the socioeconomic conditions of these communities in different countries (Qasim et al. 2010). In countries such as Afghanistan, Bangladesh, India, and Pakistan, the native medicinal flora is facing the depletion threat due to biotic interference, less rainfall, cutting of vegetation for fuel and fodder purposes, lower water table, high deforestation, accidental fires, and overgrazing. In some Asian countries, medicinal halophytes are declining because of over exploitation. Ashfaq et al. (2019) have reported the medicinal halophytes significance for primary health care among the rural communities.

According to Qureshi et al. (2007), 84% of population in Pakistan depends on traditional medicines for the cure of various health disorders. Previous surveys have described that about 60,000 traditional healers are involved in this practice in remote areas and utilize medicinal plants against several ailments (Rehman et al. 2017). Medicinal halophytes are used against genito-urinary disorders, pain, toothache, digestion problems, fever, and in respiratory and skin diseases (Qasim et al. 2010). Ghazanfar (1994) in her book on medicinal plants of Arabian Peninsula has reported several medicinal uses of halophytes. She mentions that *Aerva javanica, Portulaca oleracea*, and *Vernonia cinerea* are used against skin diseases and stings; *Artemesia* spp. as antihelminthic, *Salvadora persica* as tooth cleanser, *Centella asiatica* and *Apium graveolens* as diuretics, *Teucrium* spp. against digestive disorders; *Phoenix dactylifera* and *Cocos nucifera* are reported to show multiple beneficial uses, these are known plant species since ancient times.

The indigenous herbal medicines rely on the properties of local medicinal plants (Bartam 1995). The traditional Unani medicinal system is practiced popularly among the rural population of Asia (especially in Pakistan, India, and other Asian countries) (Obón et al. 2021). Unani medicinal system was adopted and documented by Muslim scholars in the magnificent period of Islamic civilization and they practiced it for many centuries in Asian countries (Khan et al. 2012). Examples of some medicinal halophytes used in this traditional medicine system are outlined below. In Iranian medicine, the leaves of *Atriplex leucoclada* are used as an emollient to treat cough and sore throat (Guerreiro 2018). In different parts of Pakistan, *Calotropis procera* in chest pain; *Cressa cretica* as expectorant; *Suaeda fruticosa* as diuretic, laxative and emetic; *Caesalpinia crista* in asthma, cough, headache and stomach upset; together with *Thespesia populnea* and some of the halophytes used in skin disorders (Rashid et al. 2000; Ksouri et al. 2012; Qasim et al. 2011). Also in Bahrain, *Cressa cretica* is used as anexpectorant; *Phragmites australis* in the treatment of rheumatism in Turkiye and Lebanon. In Lebanon, *Limbar crithmoides* is a well-known halophytes used as tonic (Priyashree et al. 2010; Rivera et al. 2012). The halophytes distributed in some countries representing east Mediterranean part of Asia, such as Turkiye, Syria, Lebanon, Palestine, and Jordan have been studied well in this connection and halophytes are mostly used in traditional folk medicine for urinary system disorders (Ozturk et al. 2014).

More than 100 medicinal halophytes are used in the treatment of various diseases in traditional Chinese medicine, and some of these are also used for commercial purposes (Kefu et al. 1995; Buhmann and Papenbrock 2013; Obón et al. 2021). Of these halophytes, *Nitraria tangutorum* leaves, fruits, and seeds are used as antiarrhythmic, antineuropathic, and antispasmodic agents to treat weaknesses in the spleen and stomach, including dyspepsia, neurasthenia, and common cold (Ksouri et al. 2012; Obón et al. 2021). The fruits of this plant have even been used in Northwest China to treat hypertension, abnormal menstrual bleeding and abdominal pain, and to support milk secretion during lactation (Ksouri et al. 2012; Obón et al. 2021).

Other oldest medical system in the world is Ayurvedic, which generally constitutes the widespread traditional health system applications in India (Ayush 2019; Obón et al. 2021). Extracts prepared from popular halophyte species belonging to the *Tamarix* genus in India are widely used in the Ayurvedic system for immunomodulatory and carminative (especially in children) purposes. The medicinal effect of these plants has been confirmed by clinical case studies as well (Buhmann and Papenbrock 2013). The extract obtained from *Salsola kali* is also used in liver complications (Kshirsagar et al. 2011).

4.2.2 Halophytes as a Source of Bioactive Compounds

These plants are potential natural source which could serve as a source of various bioactive molecules and innovative drugs (Stanković et al. 2015; Dajić Stevanović et al. 2019; Luković et al. 2021). Several primary and secondary compounds in the bioactive forms such as natural vitamins, fatty acids, amino acids, flavonoids, alkaloids, terpenes, isoflavonoids, saponins, anthocyanins, and phenolic acids are found in these plants (Arya et al. 2019; Luković et al. 2021). This interesting group of plant resources can play an essential role in community health and safety (Lagos et al. 2015; Arya et al. 2019). Particularly, they may provide bioactivity compunds with potential nutraceutical features and necessary for the treatment of a disease and/or for people to lead a healthy lifestyle (Kalra 2003). These bioactive substances are generally known as functional enzymes, probiotics, proteins, saponins, phytosterols, phytic acids, peptides, isoflavones, and fibers (Stankovi et al. 2015; Jdey et al. 2017; Arya et al. 2019). The halophytes are characterized by bioactive compounds and various phytochemical profiles and often show many bioactivity properties such as antioxidant, antimicrobial, and anticancer (Stanković et al. 2015). Many researchers have reported that halophyte taxa specific to the families such as Chenopodiaceae, Fabaceae, Tamaricaceae, Amaranthaceae, Plantaginaceae, Juncaceae, Plumbaginaceae, and Acanthaceae have attracted attention as a source of bioactive compounds (Dajić Stevanović et al. 2019).

4.2.3 Halophytes and Nutritional Lipids

The adequate optimal composition as well as content of lipids and other nutritive compounds in the seeds of some halophytes such as *Salicornia* species (especially *Salicornia europaea* and *Salicornia bigelovii*), *Suaeda salsa, Suaeda aralocaspica, Halocnemum strobilaceum* and *Nitraria sibirica* have revealed that these can be used as alternative oilseed sources (Geissler et al. 2013; Abideen et al. 2015; Ozturk et al. 2018g; Nanduri et al. 2019; Altay and Ozturk 2021). The seeds of species such as *Salicornia fruticosa* and *Kochia scoparia* have been reported to be the sources of fats and fat-soluble vitamins of good-quality edible oils (Geissler et al. 2013). Atia et al. (2019) and Ozturk et al. (2021b) have reported that *Portulaca oleracea*, a well-known food plant, is rich in omega-3 FA, while Nanduri et al. (2019) mention that Quinoa (*Chenopodium quinoa*), known as a promising pseudocereal, irrigated with salt water, contains high levels of unsaturated fatty acids, proteins, fiber, and minerals in terms of oleic and linoleic acids.

4.2.4 Halophytes as a Source of Food and Nutrition

The conventional crops are difficult to grow in saline regions which could develop the gap between demand and supply of good quality food in saline regions. Halophytes are the potential source of nutrition, thereby securing the livelihood of people in saline areas (Ganesan et al. 2019). Despite all this information, the halophytes are considered as underutilized sources (Lieth and Hamdy 1999). They possess great significance for subsistence and can play big role in local cultural heritage. *Beta vulgaris* and *Phoenix dactylifera* are high salt-tolerant foods eaten by people in Asia. Leaf and stem of *Atriplex halimus, Portulaca oleracea, Sesuvium portulacastrum, Chenopodium album, Salicornia bigelovii, Suaeda maritima* and *Atriplex hortensis* have been and are consumed even now as vegetables and salads in many countries (Khan and Qaiser 2006; Ozturk et al. 2008, 2014, 2016, 2018g, 2021b; Altay and Ozturk 2021). *Suaeda fruticosa, Salicornia* spp., and *Halogeton* spp. are economically important halophyte taxa as the sources of high-quality edible oil

(Weber et al. 2007; Ozturk et al. 2018g; Altay and Ozturk 2021). Khan and Qaiser (2006) have provided a checklist of 410 salt-tolerant plants from Pakistan and reported significant economic uses of 274 salt-tolerant species. In Southwest Asia evaluation of halophytes as food is as follows: the Mediterranean and Arabian Gulf countries use over 10 taxa, as against the 40 taxa consumed in Afghanistan, Iran, Iraq, and Pakistan (Ozturk et al. 2019a). Halophytes' economic usage is one of the most advantageous options for saline soil and saline water economic utilization (Kokab and Ahmad 2010).

Although not widely used on a global scale, halophytes are generally used as food by local people, mainly as vegetables, salads, and pickles (Ozturk et al. 2019a, b, c, 2021a). Among these halophytes those worth mentioning are *Salicornia, Portulaca, Sesuvium, Sporobolus, Chenopodium,* and *Suaeda* genera (Ozturk et al. 2019a, 2021b; Altay and Ozturk 2021). The halophytes are at the same time most important candidates for future use both as fresh as well as processed food, due to their functional and health properties (Loconsole et al. 2019; Luković et al. 2021).

Some halophyte plants, such as date palm (especially fruits) have been domesticated and cultivated for thousands of years and are widely consumed as food. There are some halophyte taxa such as *Crithmum, Salicornia,* and *Portulaca*, which are collected from nature and consumed with great admiration in traditional cuisines, and some like *Suaeda* and/or *Arthrocnemum* are usually collected and consumed by localas in times of famine. Under all circumstances, the halophytes have started taking place as an alternative food source in gastronomic cuisine (Ríos et al. 2021).

The edible parts of *Sesuvium portulacastrum* (especially stem and leaves) are rich in minerals such as Ca, Fe, and carotene, suitable for vegetable consumption and traditionally consumed as a vegetable by local people in arid regions of India; the leaves of some *Salicornia* species, especially *Salicornia bigelovii*, are rich in omega-3 polyunsaturated fatty acids, making it advantageous to consume these plants as vegetables (Lokhande et al. 2009; Zerai et al. 2010; Bhat and Hakeem 2021).

The use of halophytes as food may be a potential commercial option for the agricultural sector, although only locally since most of the agricultural crops show low salt tolerance and these may become unnecessary in the future as the soil salinity increases with climate change. The plants from this interesting ecological group are regarded as nontraditional products, although not easily available on our dining table due to our established food preferences, their salt-tolerance may provide the potential for their use as human food in the future (Shrivastava and Kumar 2015; Bhat and Hakeem 2021; Ozturk et al. 2021b).

4.2.5 Halophytes as a Source of Animal Feed and Forage

In arid and semi-arid areas, feed reserves are very important for livestock, which is one of the most important source of income for the localals who generally depend on the breeding of camels, sheep, goats and other herbivores. In these areas, halophyte plants are often preferred as natural feed reserves. The salt-tolerant characteristics of halophytes provide a potential for their use as livestock fodder in the future. Among these plants, *Atriplex* spp., *Nitraria retusa* and *Salsola* spp. are attracting much attention (Khattab 2007; El Shaer 2021). The species of genera *Salsola* and *Suaeda* generally flourish on saline habitats as well as under extreme climatic conditions, these can serve as a promising animal feed source. The species of these genera are of significant importance and many taxa can be used as animal fodder (Altay and Ozturk 2021).

A recently published study reports nearly 331 fodder halophyte taxa with distribution in Southwest Asia (Ozturk et al. 2019a). In the Arabian Gulf countries, *Haloxylon salicornicum, Convolvulus glomeratus, Blepharis ciliaris, Cleome brachycarpa, Senna italica, Tecomella undulata, Zaleya pentandra, Leptadenia pyrotechnica, Lycium shawii, Aerva javanica,* and *Aizoon canariense* are serving as a good fodder for camels, cattle, goats, and sheep as major halophyes, while *Halocnemum, Salsola, Suaeda, Sarcocornia, Haloxylon, Aeluropus,* and *Aellenia* are among the 20 taxa evaluated as animal feed in the Mediterranean part of Southwest Asia. In Afghanistan, Iran, Iraq, and Pakistan the number of halophytes used as animal feed exceeds 100 taxa (Ozturk et al. 2019a). Evaluation of necessary information on the ecological and ecophysiological status of halophytes naturally distributed in arid and semi-arid areas sheds light on the importance of the establishment of sustainable forage and livestock production in saline grassland habitats and develop appropriate management practices (Dajic Stevanovic et al. 2008; Luković et al. 2021). As an example, feeding with *Panicum turgidum* a halophyte is reported to have resulted in better meat quality in the livestock (Abideen et al. 2011). The quality of forage from halophytes varies a lot as it is affected by various factors such as palatability, feeding value, nutrient digestibility, nutrient composition, quantity of plant secondary metabolites, and voluntary animal feeding. The analysis of nutrient contents and chemical composition of the feed can be determined and the forage quality as well as suitability for cattle put forward (Duncan et al. 2000; Ahmad et al. 2021).

Biomass production and nutritional values of halophytes in arid and semi-arid areas may change from region to region, depending on the ecological conditions of the area, as well as from season to season (Le Houérou 1992; El Shaer 1997, 2021). This situation may lead to the variation in the flavor and nutritional values of existing halophytes (Le Houérou 1992; El Shaer 1997, 2021). Many studies report that many halophytic plants contain sufficient levels of crude protein and other essential nutrients to meet the nutritional requirements of animals under suitable climatic conditions (especially during the rainy seasons) (Kandil and El Shaer 1988; Arieli et al. 1989; Le Houérou 1992; Khan and Qaiser 2006; El Shaer 2010, 2021). Halophytes have potential animal feed properties due to the complementary crude proteins they contain, especially under arid conditions, as well as in times of famine. The species belonging to the genus *Atriplex* have been evaluated as an alternative feed source. Riasi et al. (2008) have emphasized that the amount of digestible "crude protein" associated with *Atriplex* may contain more crude protein than many other meadow plants. A similar study has been published by El-Hack et al. (2018). The halophytes can produce fodder for livestock, especially in the areas where livestock is very limited because of fodder shortage and is directly linked to human survival (Khan 2003, 2016). There is a preference for choosing different halophytic plant species by different animals. They prefer to consume different species of these plants. For example, *Alhagi maurorum* and *Atriplex leucoclada* are preferred by sheep and goats, while camels prefer to eat *Salicornia fruticosa* (Glenn et al. 1999; Ahmad et al. 2021).

4.2.6 Cash Crop Halophytes as Potential Genetic Resources

Parallel to the global population growth, the problem of food demand is constantly increasing with a decrease in the agricultural lands. To overcome this problem, alternative/complementary plant species that can be included in existing agricultural systems are investigated. The halophytes have come to the forefront recently as edible wild plant species. These have emerged as a promising option very recently as a potential genetic resource (Ladeiro 2012; Molina et al. 2014; Corrêa et al. 2021). In fact, many uses have been proposed for these plants, including food and/or feed production as well as industrial raw materials and chemicals, for landscaping, plant breeding, and evaluation for health benefits (Koyro et al. 2011; Ventura et al. 2014; Petropoulos et al. 2018, 2019; Corrêa et al. 2021).

This special group of plants occupies a special place in ecological and ecophysiological terms. These can be used to save and rehabilitate lands that are currently uncultivated and will not generate income, as well as make important contributions to potentially sustainable farming systems, although on a small scale (Ventura and Sagi 2013; Corrêa et al. 2021). Some important abiotic factors such as salinity and drought pose great challenges to food security and sustainable agriculture (Murdiyarso 2000; Ahmad et al. 2021). On a global scale, these factors can cause a significant decrease in crop yields, as well as adversely affect food and animal populations (Ahmad et al. 2021). The fact that most of the cereals with high economic value are sensitive to salt, negatively affecting crop productivity, poses a major problem in terms of food safety (Ventura and Sagi 2013; Ahmad et al. 2021). Therefore, the high potential for survival of halophytes in saline habitats provides excellent opportunities for their cultivation as potential food and forage crops (Colmer et al. 2006; Ahmad et al. 2021). One of the potential strategies to ensure food security could be halophyte cultivation in arid and semi-arid regions (Ahmad et al. 2021). According to Koyro et al. (2013), introducing the concept of sustainable agriculture could be a good solution for future food crisis. The so-called "cash crop halophytes" can prevent the waste of limited freshwater resources and overcome the water reserves problems in crop growing. Many halophyte plant taxa possess a potential commercial value as "cash crops" (Lieth and Menzel 1999). The utilization of halophytes for industrial, ecological, or agricultural purposes is possible according to the data published today. Some halophytes are used as food even at present and some for medicinal purposes, some as forage, and oil seed crops in agronomic field trials (Li et al. 2019). Eisa et al. (2017) *Chenopodium quinoa* is accepted as an alternative cash crop worth growing in the salt-affected soils of Egypt and very promising results have been obtained in terms of quality of the produced seeds. Similarly *Aster tripolium* too is considered a potential candidate for biosaline agriculture as it shows improved salt tolerance through filtering excessive Na and Cl ions in lateral shoots as well as a positive response to elevated CO_2 through increasing photosynthetic rate and water use efficiency (Geissler et al. 2009). Currently, more research is needed to fine-tune all the requirements for sustainable production and rational use of halophytes as natural resources, the cultivation practices for such species are based on practices for related conventional crops (Corrêa et al. 2021).

4.2.7 Halophytes as a Source of Bioenergy

These plants are of great significance as a source of bioenergy because salt tolerant crops can be given sea water which does not cause any reduction in seed yield and biomass (Ozturk et al. 2017d). Recently several halophytic plants have been identified having potential to produce large biomass on salt effected lands (Ventura et al. 2014). Xian-Zhao et al. (2012) mention about the halophytes whose use has been banned for the production of biofuel in China. *Atriplex, Saueda, Salicornia,* and *Distichlus* spp. are the halophytes that have high lignocellulose content. These can be evaluated for biofuel production as well as improve the soil quality and reduce the environmental footprint. The National Aeronautics and Space administration (NASA) have *Rhizospora mangle* and *Avicenia berminans* in green lab as a source of alternative energy (Bomani et al. 2011).

According to Ozturk et al. (2017d) and Bhat and Hakeem (2021), the use of green wealth “feedstock” for biofuel generation can only be made in practice if we can recognize better alternative plants, able to flourish in saline environs and must show least competition for conservative agricultural services. Halophytes have an incredible capacity to flourish in high saline environments. If these plants get adapted in such environments they can be an excellent option as “bioenergy” resource. Both oils extracted from the seeds and the “lignocellulosic biomass” from halophytes can be used for biofuel extraction. With the presence of high content of “lignocellulosic biomass” and their potential standing in high salt environments, halophytes are considered as an alternative and a valuable candidate for the production of biofuels.

The nonconventional plants or halophytes can produce two valuable harvests the seed oil and lignocellulosic biomass and straw to produce biofuel separately. These plants produce high oil yield, the oil containing a huge amount of long-chain fatty acids (Joshi et al. 2021). The oil yield and fatty acids composition varies in differing species. In the case of *Suaeda aralocaspica,* yield is more than 25% oil, as against *Suaeda acuminata* yield of only 14%. The *Crithmum maritimum* seeds contain up to 45% oil dominated by oleic acid, while *Salicornia brachiata* seeds contain 35% oil but it is dominated by linoleic acid. The oilseed halophyte oil production is reported to be potentially useful to produced biofuel, mainly biodiesel. The biodiesel is primarily composed of monoalkyl esters of long-chain fatty acids where their configuration provides reaction sites for the addition or cracking of the functional group to produce liquid fuel with miscellaneous features (Kinder and Rahmes 2009; Joshi et al. 2021).

The cultivation of halophytes for the production of biodiesel could be a reasonable option for sustainable use of saline lands, as these plants are well adapted to the salinity and can tolerate irrigation with saline water. *Salicornia bigelovii* is cultivated for biodiesel and oilseed production on farms in Mexico (Luković et al. 2021). *Aster tripolium* too is recognized as a potential cash crop halophyte, with higher content of triacylglycerol compounds in seeds than *Suaeda fruticosa* and *Chenopodium quinoa*. Different species of *Salicornia* are considered as potential new crops for biodiesel production, but there are some drawbacks such as a lower yield of biomass (Duarte et al. 2019; Luković et al. 2021). Many studies have enlightened the fact that the "fatty acid methyl ester" composition of oils produced from the plants growing on saline habitats is as good as the extraction of biodiesel from other plants (Abideen et al. 2012; Gul et al. 2013). Currently studies are continuing on the possible capacity under saline conditions and these has been identified to yield large amounts of precious metabolites, which possibly can be converted to produce oil as fuel (Hastilestari et al. 2013; Bhat and Hakeem 2021).

4.2.8 Halophytes as Landscaping and Ornamental Plants

Ornamental plants used in gardening, landscaping and as cut flowers are products of high economic value on global scale (Altay 2012; Azadi et al. 2016; Kishi-Kaboshi et al. 2018; García-Caparrós and Lao 2018; García-Caparrós et al. 2021). Halophytes can be suggested as an environmentally and sustainable alternative choice for the restoration of urban and degraded areas (Ozturk et al. 2014; García-Caparrós et al. 2021). These plants have a high survival rate in different stressed habitats due to their ecological abilities (García-Caparrós et al. 2021). The intensity of anthropogenic activities is increasing as an increase in artificial structures such as industrial areas, urban areas, roads and buildings increases, and the idea that new healthy green areas with aesthetic qualities should be created as an urgent priority (Poortinga et al. 2011; García-Caparrós et al. 2021). Although the application of ornamental plants is considered a good alternative, especially in urban areas, it is necessary to have correct ecological information about the plant species to be planted (Ferrante et al. 2011; Farieri et al. 2016). In this sense, the evaluation of halophytes as ornamental plants can be very practical (Ferrante et al. 2011).

During the last three decades, the surface of salt marshes has got reduced because of the high anthropogenic pressures originating from agriculture and the high demand for tourist urban facilities (Litalien and Zeeb 2020). Despite all this, the implementation of local halophytes in these areas could be a promising alternative, maybe the introduction of biosaline agricultural approach with the aim of restoring biodiversity and productivity could be another environmentally sustainable choice (Daoud et al. 2016; García-Caparrós et al. 2021). The ornamentals used in the restoration of urban areas initially prefer because of their attractiveness, flower colors or fragrances. Unfortunately, the maintenance required for these plants is very costly, and irrigation with poor quality water damages both the sensitive and aesthetic values of these plants thereby not accepted as a sustainable environmental approach (Ferguson and Grattan 2005; Grahn and Stigsdotter 2010). Therefore, for urban areas restoration, use of halophytes can be considered an environmentally sustainable option (Miyamoto et al. 2001; Miyamoto and White 2002). Moreover, in the case of dry ecosystems, marginal lands, salt marshes and even urban areas the restoration of roads and roadsides is of particular importance among ecologists and government agencies (Celentano and Rousseau 2016; García-Caparrós et al. 2021).

The most common uses of ornamental plants include professional landscaping (especially in urban habitats), home gardening, and cut flowers (Altay et al. 2010a, 2010b; Osma et al. 2010; Dobres 2011; Eskin et al. 2012; Tarakçı et al. 2012; García-Caparrós et al. 2021). The native species in general are well adapted to stressed conditions. The increase in soil salinization and due to scarcity of freshwater, which is increasing day by day, take us towards the most viable solution, that is, the utilization of halophytes for landscaping (Ozturk et al. 2014). Cassaniti et al. (2013) points out that the ornamental use of halophytes is a must; they have provided an extensive list of different botanical families as well as species for their use in ornamental landscaping. Much work has been done lately on the halophytes in the east Mediterranean. Ozturk et al. (2014) have reported that *Ipomoea imperati, Salsola tetrandra*, *Halimione portulacoides, Inula crithmoides, Suaeda fruticose,* and *Arundo donax* can be used as ornamentals.

4.2.9 Halophytes and Climate Change Mitigation Potential

Within the scope of the global warming scenario, it is estimated that the emergence of severe drought in many areas will cause the expansion of arid and semi-arid areas. As a result of these effects, these may cause an increase in freshwater scarcity and salinization of the existing lands. Over time, these may lead to a decrease in cultivated areas as well as the yield. This will endanger food security on a global scale (Wang et al. 2003; Imanberdieva et al. 2018b; Luković et al. 2021). Keeping this in view, the halophytes have potential to tolerate abiotic factors such as salinity and drought, can easily cope with the effects of global climate change. This situation offers a promising alternative as farming option for these plants in future. Moreover, United Nations Environment Program (UNEP) (1993) has suggested that the use of "Cash Crop Halophytes" and the ability of these plants to capture CO_2 could serve as a potential alternative natural resource to reduce the negative impacts on agriculture and the environment caused by global climate change (Caçador et al. 2002; Geissler et al. 2013; Luković et al. 2021). The vegetation of saline habitats is of great significance as it can prevent soil erosion, replenish sodic and saline soils which are imperative in the present global climate change scenario. In developed countries, there is a focus on policymaking and research on the crops that can grow well under limited resources. Mangroves and halophytes are looked upon as the alternative crops for the next generation because they can give sustainable yields under varying environments. *Suaeda fruticosa* is the best example of mitigation potential in changing climatic conditions because it can complete its life cycle under the salinity of 65 ds m^{-1} (Wungrampha et al. 2019).

4.2.10 Halophytes as Carbon Sequestration Potential

For C sequestration in arid and semi-arid regions a large-scale production of halophytes might be an attractive alternative. The reason being that salt-tolerant species can be grown on lands which is usually too saline and unfavorable for the production of most crops and trees. A high rate of productivity is being recorded from the salrt marshes and these areas are excellent sinks for C fixation. Their carbon sequestration capacity is higher than other wetland systems. They show good potential to sequester carbon continuously over thousands of years (Caçador et al. 2016).

Glenn et al. (1993, 1999) have studied halophytes as biomass crops used for the sequestration of 0.7 Gt C. The plants from this group play an important role similar to other plants in the environment by absorbing C from the air. Their ability in carbon sequestration can enrich the status of organic carbon in any infertile soil. These authors have also reported that some halophytes are proposed as promising candidates for carbon sequestration, referring to species of genera *Atriplex, Batis, Salicornia, Suaeda*, and *Sesuvium*. The studies conducted by ICAR-CSSRI research groups in India have reported many species with high potential for C sequestration such as *Populus deltoids* and *Eucalyptus tereticornis* (Chinchmalatpure et al. 2015).

4.2.11 Halophytes as Potential of Phytoremediation

Halophytes have great potential for phytoremediation of heavy metals present in saline soils, the reason being their high tolerance toward heavy metals (Liang et al. 2017; Caparrós et al. 2022). Many researchers have conducted research on the halophytes as the potential for phytoremediation of heavy metals present in the soils, indicating the environmental significance of halophytes (Christofilopoulus et al. 2016; Santos et al. 2015; Caparrós et al. 2022). *Atriplex halimus* and *Tamarix africana* are well-known examples of halophytes' potential for phytoremediation (Manousaki and Kalogerakis 2009; Santos et al. 2017; Caparrós et al. 2022).

4.2.12 Halophytes: Potential Economic Resources as Cultural Values

The saline habitats represent a complex ecological and socioeconomic system. They create an environment for a huge number of plants and animals; also offering a multitude of services of cultural importance (Bingzhen et al. 2018). The halophyte vegetation has a potential to provide cultural services which humans try to get from other ecosystems such as recreation, tourism and aesthetic values, intellectual development, and spiritual enrichment together with social relations for specific interest groups. The cultural services of higher priority include the aesthetic value, recreation and tourism, cultural heritage, as well as social relations for specific interest groups (Everard et al. 2010) such as herbal tours for observing rare and endemic plants (botanical safari) and bird watching (Everard et al. 2010).

A large number of visitors are attracted for enjoying natural saline habitats with outstanding landscapes. The saline habitats provide many opportunities for recreational activities, through their aesthetic values. One can walk, hike, go cycling, and study nature. The ecotourism in such habitats will lead towards an increase in the number of visitors, trying to follow recreation in natural areas. The cultural ecosystem services in such areas will include the benefits to visitors and income opportunities for nature tourism service providers (Luković et al. 2021). Nature-based tourism is reflected in visits and exploration of unusual saline flora and vegetation through herbal, educational, or photo tours (Loconsole et al. 2019; Luković et al. 2021). Keeping in the mind that most of these habitats are protected areas, their utilization level is low. The services they provide are mainly regulatory and less cultural. In addition to scientific and educational activities, cultural-artistic activities (painting colonies) and touristic visits can be organized in such habitats, so as to learn more about rare plant species distributed here or to observe birds (Luković et al. 2021).

4.2.13 Other Potential Economic Uses of Halophytes

There are some provisioning services, which might be considered here. These refer to the halophytes as a source of fibers, use as shelter and construction material. Some salt marsh plants (particularly *Bolboschoenus maritimus, Phragmites communis, P. karka*, and *Typha latifolia*) are the sources used for construction purposes (Luković et al. 2021). The saline habit plant sources can be evaluated as biomass and used in many industrial applications. Abideen et al. (2011) have pointed out that saline habite plants are one of the most productive sources of lignocellulosic biomass. Very high amounts of cellulose are found in the xerohalophytic plants such as *Suaeda fruticosa* and *Atriplex halimus*, while *Limonium boitardii* is rich in lignin. In the halophytes lignocellulosic biomass can be considered as material for future applications in the production of good quality paper and reinforcement of polymers. The saline steppes, particularly the saline grasslands, are dominated by xerohalophytic plant taxa which can be used as renewable resources by different fiber industries as an example; *Suaeda fruticosa* could be used in the production of high-quality lightweight paper (Salem et al. 2019; Luković et al. 2021). Moreover, on the physical and chemical characteristics and composition we find good thermal and acoustic insulation, good mechanical features, low density, high specific stiffness, biodegradability, eco-friendliness, and toxicological harmlessness as the characteristics of fiber plants where halophytes can be used. A range of halophytes represent great potential for use in plastic industry (as reinforcement in polymers) as well as automotive, and packaging industries (Salem et al. 2019; Luković et al. 2021).

References

Abbasi AM, Khan M, Ahmad M et al (2010) Ethnopharmacological application of medicinal plants to cure skin diseases and in folk cosmetics among the tribal communities of North-West Frontier Province, Pakistan. J Ethnopharmacol 128:322–335

Abebe D, Debella A, Urga K (2003) Medicinal plants and other useful plants of Ethiopia. EHNRI/Camerapix Publisher International, Nairobi, pp 188–194

Abideen Z, Ansari R, Khan MA (2011) Halophyte: potential source of lignocellulosic biomass for ethanol production. Biomass Bioenergy 35:1818–1822

Abideen Z, Ansari R, Gul B, Khan MA (2012) The place of halophytes in Pakistan's biofuel industry. Biofuels 3:211–220

Abideen Z, Qasim M, Rasheed A et al (2015) Antioxidant activity and polyphenolic content of *Phragmites karka* under saline conditions. Pak J Bot 47:813–818

Ahmad H, Waseem M (2006) Conservation status of some medicinal plants of the salt range. Zonas Áridas 8(1):40–47

Ahmad M, Qureshi R, Arshad M et al (2009) Traditional herbal remedies used for the treatment of diabetes from district Attock (Pakistan). Pak J Bot 41(6):2777–2782

Ahmad M, Sultana S, Fazl-i-Hadi S et al (2014) An ethnobotanical study of medicinal plants in high mountainous region of Chail valley (district SwatPakistan). J Ethnobiol Ethnomed 10:36

Ahmad S, Zafar M, Shinwari S et al (2020) Ethno-medicinal plants and traditional knowledge linked to primary health care among the indigenous communities living in western hilly slopes of Dera Ghazi Khan, Pakistan. Pak J Bot 52(2):1–12

Ahmad F, Hameed M, Ahmad MSA, Ashraf M (2021) Ensuring food security of arid regions through sustainable cultivation of halophytes. In: Grigore M-N (ed) Handbook of halophytes: from molecules to ecosystems towards biosaline agriculture. Springer, Cham, pp 2191–2210

Altay V (2012) Mustafa Kemal Üniversitesi Tayfur Ata Sökmen Kampüsü (Hatay)'nün süs bitkileri. Karadeniz Fen Bilimleri Dergisi 2(6):11–26

Altay V, Çelik O (2011) Antakya semt pazarlarındaki bazı doğal bitkilerin etnobotanik yönden araştırılması. Biyoloji Bilimleri Araştırma Dergisi 4(2):137–139

Altay V, Karahan F (2012) Tayfur Sökmen Kampüsü (Antakya-Hatay) ve çevresinde bulunan bitkiler üzerine etnobotanik bir araştırma. Karadeniz Fen Bilimleri Dergisi 2(7):13–28

Altay V, Karahan F (2017) Anadolu geleneksel tıbbında güneş çarpması ve güneş yanığı tedavisinde kullanılan tıbbi bitkiler. Erzincan Univ J Sci Technol 10(1):124–137

Altay V, Ozturk M (2021) The genera *Salsola* and *Suaeda* (Amaranthaceae) and their value as fodder. In: Grigore M-N (ed) Handbook of halophytes: from molecules to ecosystems towards biosaline agriculture. Springer, Cham, pp 2333–2344

Altay V, Ozyıgıt II, Yarcı C (2010a) Urban flora and ecological characteristics of the Kartal District (Istanbul): a contribution to urban ecology in Turkey. Sci Res Essays 5(2):183–200

Altay V, Ozyıgıt II, Yarcı C (2010b) Urban ecological characteristics and vascular wall flora on the Anatolian side of Istanbul, Turkey. Maejo Int J Sci Technol 4(3):483–495
Altay V, Keskin M, Karahan F (2015b) An assessment of the plant biodiversity of Mustafa Kemal University Tayfur Sokmen Campus (Hatay-Turkey) for the view of human health. Int J Sci Technol Res 1(2):83–103
Altay V, Karahan F, Sarcan YB, Ilçım A (2015a) An ethnobotanical research on wild plants sold in Kırıkhan district (Hatay/Turkey) herbalists and local markets. Biol Divers Conserv 8(2):81–91
Amjad MS, Arshad M, Saboor A et al (2017) Ethnobotanical profiling of the medicinal flora of Kotli, Azad Jammu and Kashmir, Pakistan: empirical reflections on multinomial logit specifications. Asian Pac J Trop Med 10(5):503–514
Arieli A, Naim E, Benjamin RW, Pasternak D (1989) The effect of feeding saltbush and sodium chloride on energy metabolism in sheep. Anim Prod 49:451–457
Aronsen JA (1989) Halop. A database of salt-tolerant plants of the world. In: Office of land studies. The University of Arizona, Tuscon, p 77
Arya SS, Devi S, Ram K et al (2019) Halophytes: the plants of therapeutic medicine. In: Hasanuzzaman M et al (eds) Ecophysiology, abiotic stress responses and utilization of halophytes. Springer, Singapore, pp 271–287
Ashfaq S, Ahmad M, Zafar M et al (2019) Medicinal plant biodiversity used among the rural communities of arid regions of northern Punjab. Pak Indian J Tradit Knowl 18(2):226–241
Atia A, Debez A, Rabhi M et al (2019) Salt tolerance and potential uses for saline agriculture of halophytes from the Poaceae. In: Gul B et al (eds) Sabkha ecosystems. Tasks for vegetation science. Springer, Cham, pp 223–237
Ayush (2019) Ministry of Ayush. http://ayush.gov.in/. Last accessed 19 Nov 2019
Azadi P, Bagheri H, Nalousi AM et al (2016) Current status and biotechnological advances in genetic engineering of ornamental plants. Biotechnol Adv 34:1073–1090
Balunas MJ, Kinghorn AD (2005) Drug discovery from medicinal plants. Life Sci 78(5):431–441
Bartam T (1995) Encyclopedia of herbal medicine. Dorset, Grace
Bhat RA, Hakeem KR (2021) Biomass production of various halophytes. In: Grigore M-N (ed) Handbook of halophytes: from molecules to ecosystems towards biosaline agriculture. Springer, Cham, pp 2211–2223
Bingzhen D, Lin Z, Yunfeng H et al (2018) Comparison of ecosystem services provided by grasslands with different utilization patterns in Chinas Inner Mongolia autonomous region. J Geogr Sci 28:1399–1414
Bomani BMM, Hendricks RC, Elbluk M et al (2011) NASA's green lab research facility – a guide for a self-sustainable renewable energy ecosystem. http://ntrs.nasa.gov/archive/nasa/casi.ntrs.nasa.gov/201200
Boyko H, Boyko E (1964) Principles and experiment regarding irrigation with highly saline and seawater without desalinization. Trans NY Acad Sci 26:1087–1102
Buhmann A, Papenbrock J (2013) An economic point of view of secondary compounds in halophytes. Funct Plant Biol 40(9):952–967
Caçador I, Costa L, Vale C (2002) The importance of halophytes in carbon cycling in salt marshes. In: Xiaojing L, Mengyu L (eds) Halophyte utilization and regional sustainable development of agriculture. Weather Press, Beijing, p 199
Caçador I, Duarte B, Marques JC, Sleimi N (2016) Carbon mitigation: a salt marsh ecosystem service in times of change. In: Khan MA et al (eds) Halophytes for food security in dry lands. Academic, London, pp 83–110
Calixto JB (2005) Twenty-five years of research on medicinal plants in Latin America: a personal view. J Ethnopharmacol 100(1):131–134
Caparrós PG, Ozturk M, Gul A et al (2022) Halophytes have potential as heavy metal phytoremediators: a comprehensive review. Environ Exp Bot 193:104666
Cassaniti C, Romano D, Hop MECM, Flowers TJ (2013) Growing floricultural crops with brackish water. Environ Exp Bot 92:165–175

Celentano D, Rousseau CAG (2016) Integral ecological restoration: restoring the link between human culture and nature. Ecol Restor 34:94–97
Chinchmalatpure AR, Camus D, Shukla M, Kad S (2015) Quasi equilibrium of soil carbon stock in saline Vertisols under different land use systems. ICAR-central soil salinity research institute, Karnal, India. Salinity News 21:2
Christofilopoulus S, Syranidou E, Gkavrou G et al (2016) The role of halophytes *Juncus acutus* L. in the remediation of mixed contamination in a hydroponic greenhouse experiment. J Chem Technol Biotechnol 91:1665–1674
Colmer TD, Flowers TJ, Munns R (2006) Use of wild relatives to improve salt tolerance in wheat. J Exp Bot 57:1059–1078
Corrêa RC, Di Gioia F, Ferreira IC, Petropoulos SA (2021) Halophytes for future horticulture: the case of small-scale farming in the Mediterranean Basin. In: Grigore M-N (ed) Handbook of halophytes: from molecules to ecosystems towards biosaline agriculture. Springer, Cham, pp 2367–2393
Cotton C (1996) Ethnobotany: principles and applications. Wiley, New York
Cox PA (2000) Will tribal knowledge survive the millennium? Science 287(5450):44–45
Dajic Stevanovic Z, Pecinar I, Kresović M et al (2008) Biodiversity, utilization and management of grasslands of salt affected soils in Serbia. Community Ecol 9:107–114
Dajić Stevanović Z, Stanković MS, Stanković J et al (2019) Use of halophytes as medicinal plants: phytochemical diversity and biological activity. In: Fujita M, Hasanuzzaman M, Shabala S (eds) Halophytes and climate change: adaptive mechanisms and potential uses. Wallingford, CABI International, pp 343–558
Daoud S, Elbrik K, Tachbibi N et al (2016) The potential use of halophytes for the development of marginal dry areas in Morocco. In: Khan MA et al (eds) Halophytes for food security in dry lands. Academic Press, London, pp 141–156
Dobres MS (2011) Prospects for commercialisation of transgenic ornamentals. In: Mou B, Scorza R (eds) Transgenic horticultural crops challenges and opportunities. CRC Press, Boca Raton, pp 305–316
Duarte B, Matos AR, Marques JC, Caçador I (2019) Lipids in halophytes: stress physiology relevance and potential future applications. In: Fujita M, Hasanuzzaman M, Shabala S (eds) Halophytes and climate change: adaptive mechanisms and potential uses. Wallingford, CABI International, pp 359–373
Duncan AJ, Frutos P, Young SA (2000) The effect of rumen adaptation to oxalic acid on selection of oxalic-acid-rich plants by goats. Br J Nutr 83(1):59–65
Eisa SS, Eid MA, Abd El-Samad EH et al (2017) *Chenopodium quinoa* Willd. A new cash crop halophyte for saline regions of Egypt. Aust J Crop Sci 11:343–351
El-Hack A, Samak DH, Noreldin AE et al (2018) Towards saving freshwater: halophytes as unconventional feedstuffs in livestock feed: a review. Environ Sci Pollut Res 25:14397–14406
El Shaer HM (1997) Sustainable utilization of halophytic plant species as livestock fodder in Egypt. In: Proceedings of the international conference on "Water management, salinity and pollution control towards sustainable irrigation in the Mediterranean region". Sept 22–26, 1997, Bary, pp 171–184
El Shaer HM (2010) Halophytes and salt-tolerant plants as potential forage for ruminants in the near East region. Small Rumin Res 91(1):3–12
El Shaer HM (2021) Potential use of halophytes and salt-tolerant forages as animal feed in the Arab Region: an overview. In: Grigore M-N (ed) Handbook of halophytes: from molecules to ecosystems towards biosaline agriculture. Springer, Cham, pp 2479–2498
Eskin B, Altay V, Ozyıgıt II, Serın M (2012) Urban vascular flora and ecological characteristics of the Pendik District (Istanbul-Turkey). Afr J Agric Res 7(4):629–646
Everard M, Jones L, Watts B (2010) Have we neglected the societal importance of sand dunes? An ecosystem services perspective. Aquat Conserv 20:476–487
Farieri E, Toscano S, Ferrante A, Romano D (2016) Identification of ornamental shrubs tolerant to saline aerosol for coastal urban and peri-urban greening. Urban For Urban Green 18:9–18

Farnsworth NR, Soejarto D (1991) Global importance of medicinal plants. Conserv Medicinal Plants 26:25–51
Ferguson L, Grattan SR (2005) How salinity damages citrus: osmotic effects and specific ion toxicities. HortTechnology 15:95–99
Ferrante A, Trivellini A, Malorgio F et al (2011) Effect of seawater on leaves of six plant species potentially useful for ornamental purposes in coastal areas. Sci Hortic 128:332–341
Figueiredo GM, Leitao-Filho HF, Begossi A (1993) Ethnobotany of Atlantic Forest coastal communities: diversity of plant uses in Gamboa (Itacuruçá Island, Brazil). Hum Ecol 21(4):419–430
Flowers TJ, Muscolo A (2015) Introduction to the special issue: halophytes in a changing world. AoB Plants 7:plv020
Ganesan M, Trivedi N, Gupta V et al (2019) Seaweed resources in India-current status of diversity and cultivation: prospects and challenges. Bot Mar 62:463–482
García-Caparrós P, Lao MT (2018) The effects of salt stress on ornamental plants and integrative cultivation practices. Sci Hortic 240:430–439
García-Caparrós P, Llanderal A, Lao MT (2021) Halophytes as an option for the restoration of degraded areas and landscaping. In: Grigore M-N (ed) Handbook of halophytes: from molecules to ecosystems towards biosaline agriculture. Springer, Cham, pp 2795–2810
Geissler N, Hussin S, Koyro H (2009) Interactive effects of NaCl salinity and elevated atmospheric CO_2 concentration on growth, photosynthesis, water relations and chemical composition of the potential cash crop halophyte *Aster tripolium* L. Environ Exp Bot 65:220–231
Geissler N, Lieth H, Koyro H-W (2013) Cash crop halophytes: the ecologically and economically sustainable use of naturally salt-resistant plants in the context of global changes. In: Ahmad P, Wani M (eds) Physiological mechanisms and adaptation strategies in plants under charging environment, vol 1. Springer, New York, pp 145–162
Getahun A (1976) Some common medicinal and poisonous plants used in Ethiopian folk medicine. Amare Getahun, Addis Ababa
Ghazanfar SA (1994) Handbook of Arabian medicinal plants. CRC Press, Boca Raton, p 265
Ghazanfar SA (1995) Coastal sabkhas: an analysis of the vegetation of Barr al Hikman. In: Khan MA, Ungar IA (eds) The biology of salt tolerant plants. Department of Botany/University of Karachi, Karachi, pp 277–283
Glenn E, Squires V, Olsen M, Frye R (1993) Potential for carbon sequestration in dry lands. Water Air Soil Pollut 70:341–355
Glenn EP, Brown JJ, Blumwald E (1999) Salt tolerance and crop potential of halophytes. Crit Rev Plant Sci 18(2):227–255
Grahn P, Stigsdotter UK (2010) The relation between perceived sensory dimensions of urban green space and stress restoration. Lands Urban Plann 94:264–275
Guerreiro C (2018) Chemical and biological characterization of halophyte plants with ethnopharmacological use in the Algarve coast. PhD Universidade do Algarve
Gul B, Abideen Z, Ansari R, Khan MA (2013) Halophytic biofuels revisited. Biofuels 4:575–577
Hastilestari BR, Mudersbach M, Tomala F et al (2013) *Euphorbia tirucalli* L. Comprehensive characterization of a drought tolerant plant with a potential as biofuel source. PLoS One 8:e63501
Heinrich M, Gibbons S (2001) Ethnopharmacology in drug discovery: an analysis of its role and potential contribution. J Pharm Pharmacol 53:425–432
Idrisi MS, Badola HK, Singh R (2010) Indigenous knowledge and medicinal use of plants by local communities in Rangit Valley, South Sikkim, India. NeBIO 1:34–45
Imanberdieva N, Severoğlu Z, Kurmanbekova G et al (2018a) Chapter 16: Plant diversity of Ala-Archa Nature Park in Kyrgyzstan with emphasis on its economic potential. In: Egamberdieva D, Öztürk M (eds) Vegetation of Central Asia and environs. Springer, Cham, pp 365–381
Imanberdieva N, Imankul B, Severoğlu Z et al (2018b) Chapter 15: Potential impacts of climate change on plant diversity of Sary-Chelek Biosphere Reserve in Kyrgyzstan. In: Egamberdieva D, Öztürk M (eds) Vegetation of Central Asia and environs. Springer, Cham, pp 349–364
Jdey A, Falleh H, Ben Jannet S et al (2017) Phytochemical investigation and antioxidant, antibacterial and anti-tyrosinase performances of six medicinal halophytes. South Afr J Bot 112:508–514

Jeruto P, Lukhoba C, Ouma G et al (2008) An ethnobotanical study of medicinal plants used by the Nandi people in Kenya. J Ethnopharmacol 116(2):370–376

Joshi A, Kanthaliya B, Arora J (2021) Halophytes: the nonconventional crops as source of biofuel production. In: Grigore M-N (ed) Handbook of halophytes: from molecules to ecosystems towards biosaline agriculture. Springer, Cham, pp 2451–2477

Kalra EK (2003) Nutraceutical-definition and introduction. Am Assoc Pharm Sci Pharm Sci 5:1–2

Kandil HM, El Shaer HM (1988) The utilization of *Atriplex nummularia* by goats and sheep in Sinai. In: Proceedings of the international symposium on the constraints and possibilities of ruminant production in dry subtropics, Nov 5–7, 1988, Cairo

Kassaye KD, Amberbir A, Getachew B, Mussema Y (2006) A historical overview of traditional medicine practices and policy in Ethiopia. Ethiop J Health Develop 20:127–134

Kayani S, Ahmad M, Sultana S et al (2015) Ethnobotany of medicinal plants among the communities of alpine and sub-alpine regions of Pakistan. J Ethnopharmacol 164:186–202

Kefu Z, Zi-Yi C, Shou-Jin F et al (1995) Halophytes in China. In: Khan MA, Ungar IA (eds) Biology of salt tolerant plants. Department of Botany, University of Karachi, Karachi, pp 284–293

Khan MA (2003) Halophyte seed germination: success and pitfalls. International symposium on optimal. In: Shaer E et al (eds) Resources utilization in salt-affected ecosystems in arid and semi-arid regions. Desert Research Institute, Cairo, pp 346–358

Khan MA (2016) Food and water security for dry regions: a new paradigm. In: Khan MA et al (eds) Halophytes for food security in dry lands. Elsevier, New York, pp 231–241

Khan MA, Duke NC (2001) Halophytes-a resource for the future. Wetl Ecol Manag 9(6):455–456

Khan MA, Qaiser M (2006) Halophytes of Pakistan: characteristics, distribution and potential economic usages. In: Khan MA et al (eds) Sabkha ecosystems, vol II. Springer, Dordrecht, pp 129–153

Khan MA, Ansari R, Ali H et al (2009) *Panicum turgidum*, a potentially sustainable cattle feed alternative to maize for saline areas. Agric Ecosyst Environ 129(4):542–546

Khan M, Khan MA, Mujtaba G, Hussain M (2012) Ethnobotanical study about medicinal plants of Poonch valley Azad Kashmir. J Animal Plant Sci 22:493–500

Khattab IMA (2007) Studies on halophytic forages as sheep fodder under arid and semi arid conditions in Egypt. PhD thesis, Faculty of Agriculture, Alexandria University, Egypt

Kinder JD, Rahmes T (2009). Evaluation of bio-derived synthetic paraffinic kerosene (bio-SPK). The Boeing Company Sustainable Biofuels Research & Technology Program. http://www.safug.org/assets/docs/biofuel-testing-summary.pdf

Kishi-Kaboshi M, Aida R, Sasaki K (2018) Genome engineering in ornamental plants: current status and future prospects. Plant Physiol Biochem 131:47–52

Kokab S, Ahmad S (2010). Characterizing salt tolerant plants using ecosystem and economic utilization potentials for Pakistan. In: Managing natural resources for sustaining future agriculture, vol. 2, issue 12, pp 1–20

Koyro HW, Khan MA, Lieth H (2011) Halophytic crops: a resource for the future to reduce the water crisis? Emir J Food Agric 2011:1–16

Koyro HW, Daoud S, Harrouni MC (2013) Salt response of some halophytes with potential interest in reclamation of saline soils: gas exchange, water use efficiency and defence mechanism. In: Shahid SA, Abdelfattah MA, Taha FK (eds) Developments in soil salinity assessment and reclamation. Innovative thinking and use of marginal soil and water resources in irrigated agriculture. Springer, Dordrecht, pp 523–542

Kshirsagar AD, Mohite R, Aggrawal AS, Suralkar UR (2011) Hepatoprotective medicinal plants of Ayurveda-a review. Asian J Pharm Clin Res 4(3):1–8

Ksouri R, Ksouri WM, Jallali I et al (2012) Medicinal halophytes: potent source of health promoting biomolecules with medical, nutraceutical and food applications. Crit Rev Biotechnol 32(4):289–326

Ladeiro B (2012) Saline agriculture in the 21st century: using salt contaminated resources to cope food requirements. J Bot 2012:1–7

Lagos JB, Vargas FL, de Oloveira TG et al (2015) Recent patents on the application of bioactive compounds in food: a short review. Curr Opin Food Sci 5:1–7

Letchamo W, Ozturk M, Altay V et al (2018) Chapter 2: a alternative potential natural genetic resource: Seabuckthorn [*Elaeagnus rhamnoides* (syn.: *Hippophae rhamnoides*)]. In: Ozturk M et al (eds) Global perspectives on underutilized crops. Springer, pp 25–82

Le Houérou HN (1992) The role of saltbushes (*Atriplex* spp.) in arid land rehabilitation in the Mediterranean Basin: a review. Agrofor Syst 18:107–148

Leonti M, Sticher O, Heinrich M (2002) Medicinal plants of the Popoluca, México: organoleptic properties as indigenous selection criteria. J Ethnopharmacol 81(3):307–315

Li G, Zhao P, Shao W (2019) Cash crop halophytes of China. In: Gul B et al (eds) Sabkha ecosystems. Tasks for vegetation science. Springer, Cham, pp 497–504

Liang L, Liu W, Sun Y et al (2017) Phytoremediation of heavy metal contaminated saline soils using halophytes: current progress and future perspectives. Environ Rev 25:269–281

Lieth H, Hamdy A (1999) Halophyte uses in different climates I: ecological and ecophysiological studies. In Proceedings of the 3rd seminar of the EU Concerted Action Group IC 18CT 96–0055, Florence, Italy, 20 July, 1998. Backhuys Publishers

Lieth U, Menzel U (1999) Halophyte database Vers 2. In: Lieth H, Moschenko M, Lohmann M et al (eds) Halophytes uses in different climates, ecological and ecophysiological studies. Backhuys Publishers, Leiden, pp 159–258

Litalien A, Zeeb B (2020) Curing the earth: a review of anthropogenic soil salinization and plant-based strategies for sustainable mitigation. Sci Total Environ 698:134235

Loconsole D, Cristiano G, De Lucia B (2019) Glassworts: from wild salt marsh species to sustainable edible crops. Agriculture 9:14

Lokhande VH, Nikam TD, Suprasanna P (2009) *Sesuvium portulacastrum* (L.) a promising halophyte: cultivation, utilization and distribution in India. Genet Resour Crop Evol 56:741–747

Luković M, Aćić S, Šoštarić I et al (2021) Management and ecosystem services of halophytic vegetation. In: Grigore M-N (ed) Handbook of halophytes: from molecules to ecosystems towards biosaline agriculture. Springer, Cham, pp 755–786

Malik K, Ahmad M, Ozturk M et al (2021) Herbals of Asia – prevalent diseases and their treatments. Springer

Manandhar NP (1995) A survey of medicinal plants of Jajarkot district, Nepal. J Ethnopharmacol 48(1):1–6

Manousaki E, Kalogerakis N (2009) Phytoextraction of Pb and cd by the Mediterranean saltbush (*Atriplex halimus* L.): metal uptake in relation to salinity. Environ Sci Pollut Res 16:844–854

Masoodi KZ, Amin I, Mansoor S et al (2020) Chapter 11: botanicals from the Himalayas with anticancer potential – an emphasis on Kashmir Himalayas. In: Öztürk M, Egamberdieva D, Pešić M (eds) Biodiversity and biomedicine – our future. Academic Press, pp 189–234

Miyamoto S, White JM (2002) Foliar salt damage of landscape plants induced by sprinkler irrigation. College Station: TWRI TR-1202, Texas Agric. Expt. Stat., Agric. Res. and Ext. Ctr. at El Paso, Texas Water Research Institute

Miyamoto S, White JM, Bader R, Omelas D (2001) El Paso guidelines for landscape uses of reclaimed water with elevated salinity. College Station: Texas A&M Univ. Agric. Res. Center at El Paso, Texas Coop. Ext

Molina M, Tardío J, Aceituno-Mata L et al (2014) Weeds and food diversity: natural yield assessment and future alternatives for traditionally consumed wild vegetables. J Ethnobiol 34:44–67

Murdiyarso D (2000) Adaptation to climatic variability and change: Asian perspectives on agriculture and food security. Environ Monit Assess 61:123–131

Muthu C, Ayyanar M, Raja N, Ignacimuthu S (2006) Medicinal plants used by traditional healers in Kancheepuram District of Tamil Nadu, India. J Ethnobiol Ethnomed 2:43

Nadembega P, Boussim JI, Nikiema JB et al (2011) Medicinal plants in baskoure, kourittenga province, Burkina Faso: an ethnobotanical study. J Ethnopharmacol 133(2):378–395

Nanduri KR, Hirich A, Salehi M et al (2019) Quinoa: a new crop for harsh environments. In: Gul B et al (eds) Sabkha ecosystems. Tasks for vegetation science. Springer, Cham, pp 301–333

Obón C, Rivera D, Verde A, Alcaraz F (2021) Ethnopharmacology and medicinal uses of extreme halophytes. In: Grigore M-N (ed) Handbook of halophytes: from molecules to ecosystems towards biosaline agriculture. Springer, Cham, pp 2707–2735

Osma E, Ozyıgıt II, Altay V, Serin M (2010) Urban vascular flora and ecological characteristics of Kadikoy district, Istanbul, Turkey. Maejo Int J Sci Technol 4(1):64–87

Ozturk M, Guvensen A, Gucel S (2008) Chapter 21: ecology and economic potential of halophytes: a case study from Turkey. In: Kafi M, Khan MA (eds) Crop and forage production using saline waters. NAM S & T Centre/Daya Publishing House, Delhi, pp 255–264

Ozturk M, Gucel S, Altay V, Altundağ E (2012) Alliums, an underutilized genetic resource in the East Mediterranean. In: Proceedings of the 6th IS on Edible Alliaceae. Acta Hort 969(ISHS 2012), pp 303–309

Ozturk M, Altay V, Gucel S, Guvensen A (2014) Halophytes in the East Mediterranean - their medicinal and other economical values. In: Khan MA et al (eds) Sabkha ecosystems: volume IV: cash crop halophyte and biodiversity conservation, Tasks for vegetation science, vol 47. Springer, Dordrecht, pp 247–272

Ozturk M, Altay V, Altundağ E, Gücel S (2016) Chapter 18: halophytic plant diversity of unique habitats in Turkey: salt mine caves of Çankırı and Iğdır. In: Khan MA et al (eds) Halophytes for food security in dry lands. Academic Press, Cambridge, MA, pp 291–315

Ozturk M, Altay V, Hakeem KR, Akçiçek E (2017a) Liquorice from Botany to Phytochemistry. Springer Briefs in Plant Science. https://doi.org/10.1007/978-3-319-74240-3

Ozturk M, Altay V, Gonenç TM (2017b) Chapter 24: herbal from high mountains in the East Mediterranean. In: Bhojraj S et al (eds) Drug discovery from herbs-approaches and applications. NAM S & T Centre/DAYA Publishing House, New Delhi, pp 327–367

Ozturk M, Altay V, Gucel S, Altundağ E (2017c) Chapter 5: plant diversity of the drylands in Southeast Anatolia-Turkey: role in human health and food security. In: Ansari AA, Gill SS (eds) Plant biodiversity: monitoring, assessment and conservation. CABI, Wallingford, pp 83–124

Ozturk M, Saba N, Altay V et al (2017d) Biomass and bioenergy: an overview of the development potential in Turkey and Malaysia. Renew Sust Energ Rev 79:1285–1302

Ozturk M, Altay V, Latiff A et al (2018a) Chapter 9: a comparative analysis of the medicinal pteridophytes in Turkey, Pakistan, and Malaysia. In: Ozturk M, Hakeem KR (eds) Plant and human health, vol 1. Springer, Cham, pp 349–390

Ozturk M, Altay V, Latiff A et al (2018b) Chapter 11: a comparative analysis of the medicinal plants used for diabetes mellitus in the traditional medicine in Turkey, Pakistan, and Malaysia. In: Ozturk M, Hakeem KR (eds) Plant and human health, vol 1. Springer, Cham, pp 409–461

Ozturk M, Altay V, Latiff A et al (2018c) Chapter 16: potential medicinal plants used in the hypertension in Turkey, Pakistan, and Malaysia. In: Ozturk M, Hakeem KR (eds) Plant and human health, vol 1. Springer, Cham, pp 595–618

Ozturk M, Gökler I, Altay V (2018d) Chapter 8: medicinal bryophytes distributed in Turkey. In: Ozturk M, Hakeem KR (eds) Plant and human health, vol 1. Springer, Cham, pp 323–348

Ozturk M, Altundağ E, Ibadullayeva SJ et al (2018e) A comparative analysis of medicinal and aromatic plants used in the traditional medicine of Iğdır (Turkey), Nakhchivan (Azerbaijan), and Tabriz (Iran). Pak J Bot 50(1):337–343

Ozturk M, Altay V, Altundağ E et al (2018f) Chapter 6: Herbals in Iğdır (Turkey), Nakhchivan (Azerbaijan), and Tabriz (Iran). In: Ozturk M, Hakeem KR (eds) Plant and human health, vol 1. Springer, Cham, pp 197–266

Ozturk M, Altay V, Orçen N et al (2018g) Chapter 3: a little known and a little consumed natural resource: Salicornia. In: Ozturk M et al (eds) Global perspectives on underutilized crops. Springer, Cham, pp 83–108

Ozturk M, Altay V, Güvensen A (2019a) Sustainable use of halophytic taxa as food and fodder: an important genetic resource in Southwest Asia. In: Ozturk M et al (eds) Ecophysiology, abiotic stress responses and utilization of halophytes. Springer, Singapore, pp 235–257

Ozturk M, Hakeem KR, Ashraf M, Ahmad MSA (eds) (2019b) Crop production technologies for sustainable use and conservation- physiological and molecular advances. Academic, Apple, Waretown

Ozturk M, Gucel S, Altay V et al (2019c) Clustering of halophytic species from Cyprus based on ionic contents. Phyton Int J Exp Bot 88(1):63–68

Ozturk M, Altay V, Efe R (2021a) Biodiversity, conservation and sustainability in Asia: volume I: prospects and challenges in West Asia and Caucasus. Springer, Cham. https://doi.org/10.1007/978-3-030-59928-7

Ozturk M, Altay V, Güvensen A (2021b) *Portulaca oleracea*: A vegetable from saline habitats. In: Grigore M-N (ed) Handbook of halophytes: from molecules to ecosystems towards biosaline agriculture, vol 2021. Springer, Cham, pp 2319–2332

Ozturk M, Khan SM, Altay V et al (2022) Biodiversity, conservation and sustainability in Asia: volume II: prospects and challenges in south and middle Asia. Springer, Cham. https://doi.org/10.1007/978-3-030-73943-0

Petropoulos SA, Karkanis A, Martins N, Ferreira ICFR (2018) Halophytic herbs of the Mediterranean basin: an alternative approach to health. Food Chem Toxicol 114:155–169

Petropoulos SA, Fernandes Â, Tzortzakis N et al (2019) Bioactive compounds content and antimicrobial activities of wild edible Asteraceae species of the Mediterranean flora under commercial cultivation conditions. Food Res Int 119:859–868

Pleskanovskaya SA, Mamedova MA, Ashıralıyeva MA et al (2019) Chapter 2: *Glycyrrhiza glabra* (liquorice) in Turkmenistan – medicinal and biological aspects. In: Öztürk M, Hakeem KR (eds) Plant and human health, – pharmacology and therapeutic uses, vol 3. Springer, Cham, pp 23–35

Poortinga W, Spence A, Whitmarsh L et al (2011) Uncertain climate: an investigation into public scepticism about anthropogenic climate change. Glob Environ Chang 21:1015–1024

Priyashree S, Jha S, Pattanayak SP (2010) A review on *Cressa cretica* Linn.: a halophytic plant. Pharmacogn Rev 4(8):161–166

Qasim M, Gulzar S, Shinwari ZK et al (2010) Traditional ethnobotanical uses of halophytes from hub, Balochistan. Pak J Bot 42(3):1543–1551

Qasim M, Gulzar S, Khan MA (2011) Halophytes as medicinal plants. In: Öztürk M et al (eds) Urbanisation, land use, land degradation and environment. Daya Publishing House, Delhi, pp 330–342

Qureshi RA, Gilani SA, Ghufran MA (2007) Ethnobotanical studies of plants of Mianwali district Punjab, Pakistan. Pak J Bot 39(7):2285–2290

Quamar M, Bera S (2014) Ethno-medico-botanical studies of plant resources of Hoshangabad district, Madhya Pradesh, India: retrospect and prospects. J Plant Sci Res 1(1):1–11

Radulovich R, Umanzor S (2021) Halophyte use and cultivation. In: Grigore M-N (ed) Handbook of halophytes: from molecules to ecosystems towards biosaline agriculture. Springer, Cham, pp 2517–2535

Rashid S, Iftekhar Q, Arshad M, Iqbal J (2000) Chemical composition and anti-bacterial activity of *Suaeda fruticosa* Forsk. From Cholistan, Pakistan. Pak J Biol Sci 3:348–349

Rehecho S, Uriarte-Pueyo I, Calvo J et al (2011) Ethnopharmacological survey of medicinal plants in Nor-Yauyos, a part of the landscape reserve Nor-Yauyos-Cochas, Peru. J Ethnopharmacol 133(1):75–85

Rehman MN, Ahmad M, Sultana S et al (2017) Relative popularity level of medicinal plants in Talagang, Punjab Province, Pakistan. Rev Bras Farmacogn 27(6):751–775

Riasi A, Mesgaran MD, Stern MD, Moreno MR (2008) Chemical composition, in situ ruminal degradability and post-ruminal disappearance of dry matter and crude protein from the halophytic plants *Kochia scoparia, Atriplex dimorphostegia, Suaeda arcuata* and Gamanthus gamacarpus. Anim Feed Sci Technol 141:209–219

Ríos S, Obón C, Martínez-Francés V et al (2021) Halophytes as food: gastroethnobotany of halophytes. In: Grigore M-N (ed) Handbook of halophytes: from molecules to ecosystems towards biosaline agriculture. Springer, Cham, pp 2639–2674

Rivera D, Matilla G, Obón C, Alcaraz F (2012) Plants and humans in the Near East and the Caucasus. Ancient and traditional uses of plants as food and medicine. An ethnobotanical diachronic review, 2 vols. (1: The landscapes. The plants: Ferns and gymnosperms; 2: The plants: Angiosperms). Murcia. Editum

Rozema J, Flowers T (2008) Crops for a salinized world. Science 322(5907):1478–1480

Salem S, Nasri S, Abidi S et al (2019) Lignocellulosic biomass from Sabkha native vegetation: a new potential source for fiberbased bioenergy and bio-materials. In: Gul B et al (eds) Sabkha ecosystems. Tasks for vegetation science. Springer, Cham, pp 407–412

Santos MSS, Pedro CA, Goncalves SC, Ferreira SMF (2015) Phytoremediation of cadmium by the facultative halophyte plant *Bolboschoenus maritimus* (L.) Palla, at different salinities. Environ Sci Pollut Res 22:15598–15609

Santos ES, Abreu MM, Peres S et al (2017) Potential of *Tamarix africana* and other halophyte species for phytostabilisation of contaminated salt marsh soils. J Soil Sediments 17:1459–1473

Sanz-Biset J, Campos-de-la-Cruz J, Epiquién-Rivera MA, Canigueral S (2009) A first survey on the medicinal plants of the Chazuta valley (Peruvian Amazon). J Ethnopharmacol 122(2):333–362

Shinwari ZK, Gilani SS (2003) Sustainable harvest of medicinal plants at Bulashbar Nullah, Astore (northern Pakistan). J Ethnopharmacol 84(2–3):289–298

Shrivastava P, Kumar R (2015) Soil salinity: a serious environmental issue and plant growth promoting bacteria as one of the tools for its alleviation. Saudi J Biol Sci 22(2):123–131

Song J, Fan H, Zhao Y et al (2008) Effect of salinity on germination, seedling emergence, seedling growth and ion accumulation of a euhalophyte *Suaeda salsa* in an intertidal zone and on saline inland. Aquat Bot 88(4):331–337

Stankovi M, Petrovi M, Godjevac D, Stevanovi ZD (2015) Screening inland halophytes from the Central Balkan for their antioxidant activity in relation to total phenolic compounds and flavonoids: are there any prospective medicinal plants. J Arid Environ 120:26–32

Stanković M, Petrović M, Gođevac D, Dajić Stevanović Z (2015) Screening inland halophytes from the Central Balkan for their antioxidant activity in relation to total phenolic compounds and flavonoids: are there any prospective medicinal plants? J Arid Environ 120:26–32

Svarstad H, Dhillion SS (2000) Responding to bioprospecting. From biodiversity in the south to medicines in the north. Spartacus Forlag

Tabuti J, Dhillion S, Lye K (2003) Traditional medicine in Bulamogi county, Uganda: its practitioners, users and viability. J Ethnopharmacol 85(1):119–129

Tarakçı S, Altay V, Keskin M, Sümer S (2012) Beykoz ve çevresi (Istanbul)'nin kent florası. Karadeniz Fen Bilimleri Dergisi 2(7):47–66

Tripathi J, Singh R, Ahirwar RP (2017) Ethnomedicinal study of plants used by tribal person for diarrhoea diseases in Tikamgarh District M.P. J Med Plants Stud 5:248–253

Ventura Y, Sagi M (2013) Halophyte crop cultivation: the case for *Salicornia* and *Sarcocornia*. Environ Exp Bot 92:144–153

Ventura Y, Eshel A, Pasternak D, Sagi M (2014) The development of halophyte-based agriculture: past and present. Ann Bot 115:529–540

Vitalini S, Iriti M, Puricelli C et al (2013) Traditional knowledge on medicinal and food plants used in Val San Giacomo (Sondrio, Italy)-an alpine ethnobotanical study. J Ethnopharmacol 145(2):517–529

Walter C, Shinwari ZK, Afzal I, Malik RN (2011) Antibacterial activity in herbal products used in Pakistan. Pak J Bot 43:155–162

Wang W, Vinocur B, Altman A (2003) Plant responses to drought, salinity and extreme temperatures: towards genetic engineering for stress tolerance. Planta 218:1–14

Weber D, Ansari R, Gul B, Khan MA (2007) Potential of halophytes as source of edible oil. J Arid Environ 68(2):315–321

Wungrampha S, Joshi R, Rathore RS et al (2019) CO_2 uptake and cholorphyll a fluorescence of *Suaeda fruticosa* grown under diurnal rhythm and after transfer to continuous dark. Photosynthesis Res 142:211

Xian-Zhao L, Chun-zhi W, Qing S, Chao-kui L (2012) The potential resource of halophytes for developing bio-energy in China zone. J Agric Food Sci Res 1:44–51
Yaseen G, Ahmad M, Sultana S et al (2015) Ethnobotany of medicinal plants in the Thar Desert (Sindh) of Pakistan. J Ethnopharmacol 163:43–59
Yensen NP (2006) Halophyte uses for the twenty-first century. In: Khan MA, Weber DJ (eds) Ecophysiology of high salinity tolerant plants. Springer, Dordrecht, pp 367–397
Younessi-Hamzekhanlu M, Ozturk M, Altay V et al (2020) Ethnopharmacological study of medicinal plants from Khoy city of West Azerbaijan-Iran. Indian J Tradit Knowl 19(2):251–267
Zerai DB, Glenn EP, Chatervedi R et al (2010) Potential for the improvement of *Salicornia bigelovii* through selective breeding. Ecol Eng 36:730–739

Chapter 5
Concluding Remarks and Future Prospects

5.1 Conclusion

Salinized lands are considered a major problem in arid and semi-arid countries. About one-third of the irrigated land of the world is salt-affected because of unsustainable irrigation approaches. Fresh water resources are depleting rapidly. In this situation, it is difficult to develop non-conventional agricultural technology that could use saline soils effectively. In this scenario, brackish or seawater utilization for agriculture in coastal and inland sandy soils are an attractive option. Such applications could be a source of feed, food, fiber, and oil on non-productive arid saline lands. Desert land is about 43% of the total earth area. About 15% of this land is close to the sea and it could be utilized for crop production.

The planting of halophytes is the key to rehabilitating salt-affected soils. This will have sound economic and social impacts on the humans living around the salinized areas. This approach is beneficial for underdeveloped arid countries where farmers have limited resources and communities are experiencing migration and poverty. Our past is a good evidence that farmers always shift from less to more tolerant crops due to environmental and other pressures including salinity. The saline habitat plants have a wide range of sustainable uses as "halophytes" while more potential taxa are still under-explored. Crop diversification on saline soils and water is the best alternative for high-income ventures. A successful cultivation of halophytes is an open unexpected opportunity for farmers. It is also a challenge for researchers to identify cash crop halophytes. The evidence of land degradation and global climate change due to salinization are well known now. It is difficult to convert saline lands into cultivated ones to meet the demand for nutritional and food security of the increasing population. Keeping this in view, halophytes are expected to play an important role in maintaining ecological and economic balance.

M. Ozturk et al., *Halophyte Plant Diversity and Public Health*,
https://doi.org/10.1007/978-3-031-21944-3_5

Asian countries show a rich vegetation on saline lands. These can reduce the gap between supply and demand for food, fodder, and economically significant other products. The potential of halophytes as a renewable source of energy can prove of great help in the mitigation of climate change as well as a reduction in the carbon footprint, as it needs to be conceptualized on the basis of the field.

5.2 Future Prospects in the Use of Halophytes

Saline habitats are one among the arid and semi-arid habitats characteristic of different climatic regions of the world. We come across a great diversity of varying saline habitats with specific ecological characteristics. These areas provide multifunctional ecosystem services including different economic and regulatory ecosystem services as well as provision and cultural services. All these have practical benefits which are derived from their sustainable use and management. Therefore, taking into account the potential adverse anthropogenic impacts on saline habitats, an urgent attention should be given towards the development of different conservation measures and promoting sustainable use and multiple management options for each identified service (Luković et al. 2021). The halophytes possess different potentials in their economic evaluation, including at the same time a multitude of ecosystem and/or biodiversity services. Thus, halophyte crops with full domestication potential need to be identified and developed urgently within the scope of viable production systems (Radulovich and Umanzor 2021).

To raise the socio-economic status of the people living around such areas, along with the sustainable cultivation, following a sustainable use and protection of this potential natural resource, many small industries can be developed by the relevant governments with small grants. Therefore, both government and private sector should urgently invest in this initiative to make halophyte plant resources a potential resource for the future (Ozturk et al. 2019a). In addition, there is an urgent need for cooperation and coordination at the local, regional, and international level (Ozturk et al. 2019a).

The following proposed steps are needed to be taken to utilize the saline lands properly.

- The coordination and cooperation at the global and regional levels are of great significance to utilize saline land and water. The involvement of institutions, farmers, politicians, and other associations is crucial because all should be familiar with their role.
- The greenhouse experiments should be performed under local climatic conditions to select the sites.
- The sustainable system of production should be developed for plantation in inland or coastal areas that will be beneficial economically.

Ozturk M, Hakeem KR, Ashraf M, Ahmad MSA (2019b) Crop production technologies for sustainable use and conservation- physiological and molecular advances. Apple Academic Press/ CRC Press

Ozturk M, Gucel S, Altay V et al (2019c) Clustering of halophytic species from Cyprus based on ionic contents. ΦYTON 88:63–68

Ozturk M, Akram NA, Turkyilmaz BU, Ashraf M (2022) Introduction and application of organic fertilizers as protectors of our environment. Cambridge Scholars Publishing, London

Radulovich R, Umanzor S (2021) Halophyte use and cultivation. In: Grigore M-N (ed) Handbook of halophytes: from molecules to ecosystems towards biosaline agriculture. Springer, Cham, pp 2517–2535

Shuyskaya EV, Rakhmankulova ZF, Toderich KN (2021) Role of proline and potassium in adaptation to salinity in different types of halophytes. In: Grigore M-N (ed) Handbook of halophytes. Springer, Cham. https://doi.org/10.1007/978-3-030-17854-3_75-1

Yasin MA, Awan AR, Anwar S et al (2021) Economic utilization of salt-affected wasteland for plant production – a case study from Pakistan. In: Grigore M-N (ed) Handbook of halophytes. Springer, Cham. https://doi.org/10.1007/978-3-030-17854-3_87-1